LOCATION
LOCATION

CONNECTIVITY

NEXT GENERATION REAL ESTATE,
INTELLIGENT INFRASTRUCTURE, AND THE
GLOBAL PLATFORM FOR COMMERCE

JAMES CARLINI

EAST DUNDEE, ILLINOIS

James Carlini/Publisher CARLINI & ASSOCIATES PRESS
445 Greenwood Avenue
East Dundee, Illinois/60118
www.carlinij.com

Book Layout ©2013 BookDesignTemplates.com

Ordering Information:
Quantity sales. Special discounts are available on quantity purchases by corporations, associations, and others. For details, contact the "Special Sales Department" at the address above.

LOCATION LOCATION CONNECTIVITY/ James Carlini. —1st ed.
ISBN 978-0-9906460-4-4

CONTENTS

v

Dedication to Holly Carlini

ACKNOWLEDGEMENTS
FOR SECOND OPINIONS

Holly Carlini, Wife & Proof-reader
Dave Bricker, Author, Florida
Kenn J. Jankowski, Long-time Executive, Colorado
Peter Rosendale, Photographer & Cover Design, Illinois
Tom Satala, retired AT&T manager and now Realtor, Florida
Art Tursh, Chicago Mercantile Exchange

[1]

PREFACE

"There are no experts in this emerging industry. The best you can be is a good student – Always Learning." – **JAMES CARLINI**

The quote above is how I started off every technology management course I taught at Northwestern University. With the diverse mix of adults and expertise in these undergraduate and graduate-level courses, we needed to quickly find common ground. From those who were novices in an emerging technology industry, to those professionals already in the field who thought they knew it all, I needed to get everyone on the same page. No one knows everything, including me. This was a good statement to start with.

What led up to writing this book? The combination of perspectives gained from advising major clients and developments on mission critical networks across four decades; two decades of teaching at Northwestern University in both undergrad and Executive Masters programs; three decades of performing expert witness work in network infrastructure technology, commercial real estate, and

Intellectual Property (IP) issues; and working as an elected official in planning, reviewing and implementing various areas of infrastructure and mission critical networks as it related to regional economic development. Add being a president of a condo board for several years and you have a good diverse mix of experiences, challenges, and actions taken to find technology solutions.

This book is needed today as we must get everyone on the same page with next-generation real estate which is dependent on technology, systems integration, and Intelligent Infrastructures. You'll learn how understanding this convergence is critical in order to compete in the economic diversity and global markets of the 21st century.

The following excerpt is from an article by Frank Bisbee, a long-time telecommunications and cabling industry pundit in Electrical Contractor magazine (August 2013). His observation provides a good starting point:

Defining new intersections of interest in the great established and evolving domains of real estate, infrastructure, technology, and regional economic development, there is an intersection where all these diverse worlds of industries, as well as complex developments, meet.

At that multifaceted crossroad, you will come across James Carlini defining the undefined. You will find him classifying and constructing the framework of interrelationships of these four vast areas in order to crystallize a multidisciplinary approach and strategy for high-tech real estate. Understanding that strategy will enable organizations to develop and shape the structure that can be used effectively in today's mobile internet age. – (http://www.ecmag.com/section/your-business/james-carlini-da-vinci-intelligent-building-movement#sthash.l44yavle.dpuf)

My book is a practitioner's strategic perspective on intelligent infrastructures, the platforms for commerce needed to provide regional economic sustainability within a 21st century global marketplace, where the three most important words in real estate are now *"Location, Location, Connectivity."*

Based on almost four decades of diversified work with real estate developers; corporate and municipal clients; systems litigation, and interviews with executives in various industries while researching global initiatives, there are definitely some new concepts many are starting to pursue and implement not only on a building basis, but also on a campus (or industrial park) basis. Along with integrated management as well as leadership principles, there are also words of wisdom to be aware of and to adopt as new guidelines as we pursue the implementation of the "Internet of Things" (IoT).

Within the book are several charts, matrices, frameworks and diagrams which help create a faster visualization of these new concepts. They are important in trying to master the global marketplace and this new high-tech real estate market. Some of these practical insights throughout the book will also be highlighted as **CARLINI-ISMS**:

CARLINI-ISM: *"These 'words of wisdom' will be throughout the book and will provide some pragmatic insights to the reader based on real experiences and not from theoretical or untested concepts."*

Many books on technology and its strategic direction are written by by-standers. They don't work in the industry and they have never gone through any multi-phase systems development process or worked on any mission critical systems integration project.

They glean off some perspectives of those they interview, read some press releases, and come up with some conclusions which are not based on first-hand experience in both successes and failures. They are susceptible to parroting those trying to hawk a specific corporate strategy or provide a very narrow view based on a single vendor and many times do not have the depth of background to even question what some are saying as it relates to a societal direction or a corporate-sponsored direction in applying technology to the organization.

MY ROOTS IN TECHNOLOGY, REAL-TIME NETWORKS, MISSION CRITICAL APPLICATIONS AND REAL ESTATE

Fortunately for you, I am no by-stander. I worked in the industry on all sides of the table, seeing many things transform and seeing some executives' strategic visions fizzle faster than their careers. I have also taken apart poorly designed systems with testimonies for civil and federal courts as well as public utility commissions.

Teaching over two thousand adult students across a couple of decades while consulting has also given me insights on how to present this material so it is easily understood and can be readily applied in a real environment.

The areas which built the groundwork for this practitioner's broad view of critical infrastructure and how concepts should be viewed from a multi-disciplinary, strategic perspective include aspects and elements of the "**RITE**" approach:

Real estate –the buildings, business campuses, and industrial parks.

Infrastructure – the supporting infrastructure including network and power along with all the transportation levels.

Technology – computers, networks, Smartphones, Tablets, wireless networks (WiFi, DAS), fiber optics, software, apps, other elements.

Economic development – regional economic (and global) sustainability and viability.

Over the last two decades, this book and its concepts have been percolating and crystallizing as work on various projects from planning to implementation as well as network infrastructure litigation have provided a rich and broad insight on what is important and what concepts stand the test of time.

This experience has built the groundwork defining a broad view of critical infrastructure and how it relates to regional economic viability. Intelligent buildings are now clustered into intelligent business campuses and other venues which require a new layer of infrastructure as well as a multi-disciplinary understanding to compete in a global economy.

Unfortunately, this is yet to become part of a university curriculum because most graduate curricula are still focused on a single discipline. These concepts and frameworks have converged disciplines and span across several industries. This is where the market has evolved and where higher education needs to evolve to, in order to prepare graduates with cognitive tools they can use.

A multi-disciplinary focus is needed to be developed and taught to next-generation graduates in the 21st century. Continuing the prior century's approach of graduating single discipline-focused experts will not address this major paradigm shift.

Concepts which have crystallized and strategies that have shaped policies which are more applicable to mainstream endeavors include:

- Solid infrastructure is key to providing a platform for economic development.
- All countries wanting more global trade are expanding their infrastructure, adding cutting-edge intelligent infrastructure (network and power) and supporting platforms for global commerce.
- Solid network platforms of broadband connectivity which support Smartphones and Tablets are being implemented in both single and multi-venue facilities to attract and maintain businesses that can employ people.
- Regional viability is key to future sustainability.

There are various perspectives that are important in shaping the way new developments should be approached. From a practitioner's perspective, who has worked on projects with developers, property management firms, and government economic development commissions in strategically applying technology to various organizations, these new ideas were learned and parlayed into practical strategic directions?

This book will give you the practical insights and wisdom of 35+ years in a dynamically changing industry where many times there were no prior roadmaps for success. You had to blaze your own trail and create your own rules-of-thumb. As I have advised clients and the thousands of students taught in these areas over the years, I will caution you with:

CARLINI-ISM: *"Just as one course in first aid does not make you a brain surgeon, one course in technology doesn't make you a Technologist."*

CARLINI-ISM: *"Document your successes as others will be sure to document your failures."*

This especially holds true when you are working with brand-new concepts which you are trying to implement in an ever-changing, technologically-dependent marketplace.

As one long-time colleague told me, *"You are becoming a 'last of a dying breed' like World War II veterans. Your experience and knowledge cannot be easily duplicated or passed down."* He is right.

Experiences focusing on quality and a sense-of-urgency for designing mission critical networks and applications will never be duplicated at one company. Hopefully, this book will encourage the reader to adopt and pursue those qualities individually. The "**CARLINI-isms**" throughout the book provide some added wisdom on leadership and everyday management along the way.

The conclusion I have arrived at and hope you will as well after reading my book is:

There is not going to be a viable economy in any city of any significance, unless they have a network infrastructure which is going to support broadband connectivity.

Enjoy the book. The concepts I have learned in the field and then taught in the classroom will become invaluable to you in adopting a broad, pragmatic, multi-disciplinary focus.

As we move into the future, we will need to invent and apply new technology in order to remain globally competitive. I hope I can inspire a new generation of thinkers who can approach problems with creative and innovative solutions based on applying technology to business and real estate.

- **JAMES CARLINI**

- **2014**
- **4712 – THE YEAR OF THE HORSE**

A CONCEPT COMES OF AGE

"Where to build our factory? "How about fantasy land?"
– RODNEY DANGERFIELD

Back in January, 1977, I started work at Bell Telephone Laboratories in Naperville, Illinois. I went to college on a full academic scholarship to be a band director, but no jobs were available at the time. Lucky for me, Western Electric (which owned half of Bell Labs) was looking for education majors who they could train extensively to get into customized software development and digital telephony engineering principles to develop real-time central and toll office software. Their new line of Electronic Switching Systems (ESS) was the computerized backbone of the Public Switched Telephone Network (PSTN) of the country. It was "the" network of the Bell System. All other corporate and enterprise voice and data networks ran through it.

Although this is not a book about Bell Labs and its wide contributions to technology, basic research, telephony, and global communications, it should be mentioned as a place which designed and built the communication and information-processing building blocks from transistors to lasers and satellites propelling the whole electronic communications infrastructure into the 21st century. More importantly, they understood how to create a work environment where you concentrated on the problems at hand, because all the other issues and minutiae were taken care of. They also spent a lot of money on furthering the education of their people which I came to value later in my career.

I feel fortunate in getting a huge education at Bell Labs at the beginning of my career when no university had their level and depth of quality network, telephony and software engineering courses. It definitely impacted my long-term perspective on real-time networks and set the foundation in focusing on quality within software engineering. They led the vast majority of corporate organizations as to the understanding, design and application of technology into the workplace.

SETTING THE STAGE: BELL SYSTEM EXPERIENCE AND EDUCATION

I learned software development and network engineering for real-time systems at a place which will never be duplicated as far as its environment to learn; develop technical expertise; foster and create innovative products; and to do things right. Those who worked at any location of Bell Labs know what I am talking about. Those who never worked at the Labs will never understand the positive atmosphere, the depth of expertise, and the reputation of the place which used to submit a patent a day for various fields of primary research as well as communications.

At that time, many people involved in software development still developed software in COBOL and JCL and got their punched cards typed up to run a program. At the Labs, everyone developed software on a terminal. Punched cards? That was already passé. You logged on and went right into "the system" to write code on-line.

"Technology was going to be part of the core business at every company" and those at Bell Labs understood that strategic insight long before most other organizations in all the industries grasped the concept.

I worked on projects utilizing the UNIX operating system long before most knew how to spell it, let alone, how to use it. In 1979, I wrote an online personnel management system for the HR department. That had to be one of the first commercial applications using UNIX and C language. UNIX and C were initially considered to be the operating system and language used in computers running the phone network. Later on, UNIX and C became very commercialized and broadly adopted as an operating system and language to develop applications across many organizations and industries.

While at Bell Labs, they sent me to a three-week course at Clemson University (1979) on the INTEL 8080 and 8086 chips that became the processor for the IBM PC and all of its clones. Bell Labs was a great environment to get a solid education on emerging technologies which were to become the future building blocks of technology for the next couple of decades.

The key concept I walked away with from Bell Labs was if you understood the technology and its benefits, you were more likely to incorporate it into the fabric of the organization's infrastructure and business applications. Getting people well-trained was important because they became more productive in a shorter period of time.

If they were productive and creative, they would be solving business issues faster than their competitive rivals.

CARLINI-ISM: *"Whoever has the best-trained workforce, is the toughest competitor."*

My career continued as I moved to Motorola and picked up background on wireless systems and police/emergency services dispatch communication systems. I did not stay there long and came back to the Bell System working at Illinois Bell headquarters in 1981.

The training I received while in the Bell System was unsurpassed by any other company, single university, or other type of training. This education included going to Bell System courses on complex telephony and computer networks at their Piscataway and Princeton, New Jersey training centers as well as graduate-level information science courses (focused on large IBM systems) at MIT in Cambridge, Massachusetts. We also attended courses in structured analysis, structured design, structured programming methodologies from several leading experts of the time.

Again, the overall objective was to get a broader understanding of the network infrastructure, software engineering, and learn more about sophisticated IBM's computer systems and networks to become more competitive. At the time, IBM had a huge footprint in just about every fortune 500 company, was a major competitor to AT&T and AT&T wanted a part of their market share. I patterned all the courses I taught later, on my education and experiences in the Bell System as well as my first-hand consulting to major corporate organizations.

CARLINI-ISM: *"The broader the perspective, the better the problem-solver, especially in applying technology to business."*

At Illinois Bell, I consulted as a strategic technical resource to American Hospital Supply Corporation (AHSC). I worked with one of their managers, Kenn J. Jankowski, on developing their strategic five-year plan for information technology and network communications across all their divisions.

AHSC, a cutting-edge company at the time, had its own private network providing all its hospital customers direct access for online ordering of equipment and supplies. In effect, it was the forerunner to the internet as anything could be ordered online from any location on their network.

What AHSC did in the late 1970s and early 1980s on a private enterprise network was what many companies wanted to do in the late 1990s and early 2000s on the internet with public access: get a direct link into the customers' ordering process on their premise.

AHSC pushed its equipment and network to the max until they had to make a strategic decision to switch from a large, distributed Burroughs (mainframe computer)/ four phase (mini-computer) environment to an IBM mainframe environment. Their success hinged on technology, network infrastructure and a well-trained cadre of it professionals who also included some network specialists.

They had an Achilles heel of their total operation: the home-grown protocol developed by one person for the Burroughs network and only he knew how everything worked.

It was the classic "single-point-of-failure" liability which people think never exists, but it does and is not uncommon.

They needed to get away from this customized protocol on their mission critical network which was making them hundreds of millions of dollars in sales each month. They needed to get a large redundant system where they could also get talent off the street to support their mission critical application.

CARLINI-ISM: *"Eliminate any single-point-of-failure on any mission critical network and/or application."*

This was one of the areas they brought me in to review, the feasibility of the switch from one vendor's hardware and network protocol to the other. Basically, the only system which could handle the amount of transactions taking place on their network was IBM. The Burroughs/Four Phase computer configuration was maxed out in its capacity to handle transactions and on busy days it would bottom out. More importantly, its communications protocol was home-grown, undocumented, and in the head of only one support person.

This was a big decision. As the second largest user of Burroughs mainframe equipment, their strategic decision to leave Burroughs affected other major users and their long-term plans. Another large Burroughs user, Harris Bank of Chicago which at the time was about eighth or ninth in the top ten Burroughs' customers, took notice and invited AHSC to some planning meetings. They did not want to wind up with a technology which would not take them far into the future.

The AHSC conversion project and strategic, long-term plan was one of the first big, diverse, technology projects I worked on. It focused on mission critical systems integration and covered a very broad area in its scope. It impacted the entire corporation's approach to business as well as its diversified divisions with their own integrated systems. I gained a lot of pragmatic insights from working on it and walked away with a much broader perspective on real-time systems integration and how different vendors had to develop ways to engineer their systems into a multi-vendor environment.

The more you work on difficult, cutting-edge projects, the more you pick up skills no one else has. Few will tackle broad-based projects and most internal corporate people will be focused more on day-to-day operations than long-term strategic systems development.

One of the problems in the industry today is the lack of in-depth training as well as the need for broad, pragmatic perspectives on what can be accomplished and what is beyond technical feasibility. Today, no one wants to make an investment in their people, yet this type of strategic investment pays off well in this area.

Being creative is also a good pre-requisite in tackling new areas of applying technology to an enterprise and that is not taught in most technical, computer courses.

GAINING A "BIG EIGHT" CONSULTATIVE APPROACH

From the pre-divestiture Bell System, Arthur Young (AY) recruited me out at the end of 1983 and hired me as their Director of Telecommunications and Computer Hardware Consulting. AY was one of the "Big Eight" accounting firms at the time, (now part of the "Big Four", Ernst & Young).

I worked on several large consulting projects focused on large real-time systems integration and applying cutting-edge network technologies to various industries. Again, my in-depth Bell System technical education proved invaluable as more companies wanted to move into network-based, mission critical applications and few people understood the public switched network as well as a quality-based systems approach in developing and integrating mission critical real-time systems.

From working with McDonald's on the concepts of in-store processors and tying them back to a headquarters system (what I coined "the McNetwork") to working on a large, factory automation project for Arco Metals at an aluminum-processing plant in Russellville, Kentucky, I worked on more strategic network and complex information technology projects across various industries. At that time, integrating multiple vendors into an enterprise network was not a simple, nor a well-established practice.

Arthur Young is where I first got involved in Intelligent Buildings and real estate developments. In 1985, I was assigned to manage a project to review the planning of a new six-building campus in Silicon Valley for Santa Fe Southern Pacific Development Company, making sure the owners got their money's worth with the technology upgrades they were paying for. At that point, I started writing trade journal articles and white papers as well. One important white paper was entitled, "Measuring a Building's IQ", published in New York University's Real Estate Review in November 1985. This was an initial document which conveyed the acknowledgement of the emerging convergence of real estate, infrastructure, and technology as well as how to measure the "intelligence" of a building which was made up of the technologies used within the building across network, information technology and building automation systems.

WIN, PLACE, OR SHOW – AND YOU GOT SHOW

Part of my AY consulting experiences was to review failed enterprise-wide computer networks or a building's network infrastructure and assess as well as report all of the flaws. Sometimes, there was friction in the room when I pointed out the fallacies of someone else's design and implementation. Few seemed to like the gunslinger approach to mission critical networks, I thrived on it. It was a challenge to see if you could accomplish a good and accurate analysis in a short period of time. It was, and still is, highly satisfying when you can go in and candidly talk about what is right and what is wrong with a complex system knowing what you say becomes the determining factor of funding a project or canning it. This also led to expert witness work later in my career.

On most projects, decision-makers were always concerned about getting their money's worth. You didn't have six months to acclimate into the organization. You might have six days to analyze the current situation and present your findings to an executive committee within two weeks of the start of the projects.

One of the most stressful executive meetings I ever sat in on throughout my whole career was at Arlington Park racetrack. The owner and the CFO sat around a long, oval table along with their lawyers and accountants all wanting to know if they were getting their money's worth on this satellite project they were funding. They were renting a satellite transponder for $64,000 a month, plus paying a consultant's fee.

The endeavor was to transmit races across the country to other racetracks during the day and selling the bandwidth to financial companies for transmission of banking transactions at night. It sounded like a great money-making idea. A "Go-No go" decision to continue funding into the next phase of this new endeavor was our objective.

The room crackled with intensity and positioned across this long conference table, three small tape recorders recorded the meeting simultaneously. I had two of my managers working with me on this short-notice project. We had to explain the three levels of satellite services to those who didn't understand technology, but were paying through the nose for it.

We only had two weeks to review this project, its Pro Forma statement, its projected revenues, and the overall capabilities of this satellite-based network. In the beginning, it sounded like a creative winner, but our analysis showed some big problems and we had to present them at this high-pressure meeting. At the time, three levels of satellite services with tier-pricing based on their availability and redundancy were the available choices. (**See CHART 2-1**)

I knew the non-technical advisors, the lawyers and accountants, as well as the CFO and owner would not understand half the technical jargon like "protected" (meaning the transponder was backed-up by another transponder on the satellite) and "unprotected" (meaning the satellite transponder had no back-up).

CHART 2-1: LEVELS OF SATELLITE SERVICES

LEVEL	Transponder CAPABILITIES	Cost per Month	LANGUAGE THE EXECS UNDERSTOOD
1	Protected, non-pre-emptible (fully redundant)	$120,000	"WIN"
2	Unprotected, non-pre-emptible (no redundancy)	$90,000	"PLACE"
3	Unprotected, pre-emptible (no redundancy, someone else's back-up)	$64,000	"SHOW"

As far as "pre-emptible" (being able to be taken over by another channel – in effect, this transponder could be someone else's back-up) and "non-pre-emptible" (meaning it could not be used for someone else's back-up), these were all terms that could not be easily described at the pace this meeting was going.

Just as I started presenting the project's feasibility analysis, it struck me. I know how to present this in "Apples and Oranges" terms. I'll use the terms they already understand and are comfortable with. So when I started to describe the complex levels of services and the needs of sophisticated banks who would be their potential customers, I said this:

"Banks and financial institutions have a lot of money riding on their transactions and they want to put their money on a "winner." That means the satellite service has to have redundancy and a back-up link. The different levels of services are like Win, Place or Show - and you have Show."

Right after I said that, you could hear a pin drop in the room and the CFO said, *"Well, I guess we have nothing further to discuss"* and he closed a big binder up and that was it. The project was finished. They never went into re-selling satellite services because it was not profitable when you added the costs of paying for a higher level of service and overlooked needed equipment (like a dish heater to melt ice on it).

CARLINI-ISM: *"Always try to explain technical terms in every-day terms and you will get a lot more positive response and support as well as clear understanding from those who do not have a technology background."*

When a system has failed or is not supporting the enterprise, no one wants to hear flowery theories, complex excuses, or glowing conceptual euphemisms something has failed. All they want to hear are the real reasons and the possible solution to the problem. (This is especially true of those who are writing the checks to pay for all the endeavors.)

The interesting conclusion to this racetrack story was it gave me the idea to teach diverse complexity of satellite services to my classes at Northwestern in the same way. "Win, Place, and Show" became the easy way to get people to remember Protected, Unprotected, Pre-Emptible and Non-Pre-Emptible service levels.

Another lesson learned was the importance to present technology in terms which could be understood by everyone around the table. That's when the best decisions could be made based on how well all the issues were presented in a clear, concise manner.

YOU'RE AN EXPERT? I MUST BE A GOD

Many times, when you are working on technology projects of this complexity, you are blazing a new trail, because there are no years of collected data, other similar projects already built-out, or historical "rules-of-thumb" already established and accepted within the industry that can be followed. You basically begin to structure the concepts from nothing. There is no "prior framework" or proven results to rely upon or refer to for industry-standard or "industry accepted rules-of-thumb."

CARLINI-ISM: *"Structuring the unstructured is a large step in the planning and design of any mission critical application."*

After you do one project, you have become an "expert" in the field. The difference between an expert and a novice in some of these accelerated endeavors is only one project when you are working on the cutting edge. Two projects make you a senior expert and three projects make you a god. That's what happens when everything converges so fast and you have to work on a project to get a handle on all the new concepts that evolve.

The rest of the book will discuss new development and planning concepts along with master planning issues, pragmatic procurement guidelines and marketing approaches in developing a platform for economic development for the 21st century.

The critical infrastructure framework, the ***Platform for Commerce***, will be discussed in detail as well as the **TARGET** Map of Technology which has been used as a conceptual model to get the point across that no matter what, "Technology And Revolutionary Gadgets Eventually Timeout."

Introducing technology and applying it into the organization is a continual process and not a one-shot endeavor. This **TARGET** model has been used as part of discussions in both undergraduate and Executive Masters programs at Northwestern University which focused on planning, designing, implementing, and managing technology within an enterprise.

Putting together frameworks and structuring concepts is a good first step in eliminating confusion and getting everyone the ability to visualize what needs to be done. Within this book there are several key frameworks and structured conceptual building blocks which can help in tackling any high-tech technology or real estate project. Learn them and use them to your advantage.

GOOD EDUCATION IS KEY TO GOOD SYSTEMS DEVELOPMENT

No company today is providing anything near the quality, nor depth, of education AT&T (the Bell System) provided. (Including the new AT&T itself)

We were focused on quality initiatives and had access to all the latest in structured design, analysis, and structured programming courses. You could say it was the *"Golden Age of Software Engineering"* in the United States, because no one else was even close to what we had.

That focus on quality has stayed with me all this time and helped shape the courses I developed for major universities, primarily Northwestern University. It also influenced the work I performed for various clients on various mission critical projects as well as lawsuits and litigation support. We need to re-emphasize that skill set and the focus on quality in today's global markets, projects, and workforces.

Quality and educating the workforce to embrace principles of doing the job right should be primary objectives in our economy today, as well as tomorrow. Quality is one of the basic foundations in infrastructure, next-generation real estate, and technology development. **(See CHART 2-2)**

If quality is not a basic underpinning in any of these areas, whatever regional economic development that is dependent on them as their foundation is compromised.

Today, too many executives have bought off on false promises of technology as well as the false promises of overnight experts touting their "ten years of expertise" on an initiative or technology that has only been around the market for two years.

These pseudo-experts do not help anyone in trying to understand how all the building blocks of technology need to be assembled. Does that sound too cynical? Unfortunately, it is the politically accurate summary of today.

CHART 2-2: QUALITY IS THE BEDROCK OF 21ST CENTURY HIGH-TECH REAL ESTATE

REGIONAL ECONOMIC DEVELOPMENT
TECHNOLOGY
INFRASTRUCTURE
REAL ESTATE
QUALITY **(The bedrock of all endeavors)**

CARLINI-ISM: *"Quality is the bedrock of all successful endeavors."*

DEVELOPING A PRAGMATIC REAL WORLD PERSPECTIVE

The best analogy I can give is the one I actually used in classes to open up the course with a short video of the BACK TO SCHOOL scene of Rodney Dangerfield sitting in the Business Professor's course and telling him how he missed all the important points when selecting the land, preparing the infrastructure, and developing space for a 30,000 square foot corporate endeavor for office space and a factory. The professor, not having any practical experience, is called out by Dangerfield who has run a big, multi-state operation.

The professor, an out-of-touch academic, omits many issues and obstacles in his cost analysis, like local zoning, delivering cement for the infrastructure platform, ongoing facility operations (like garbage collection), and price per square foot construction costs which would be naturally understood and analyzed by a practitioner in the field. He dismisses Dangerfield's comments, which are totally on target, and asks his final question to the class of, "Where to build our factory?" Rodney Dangerfield replies, "How about Fantasy Land." (http://www.youtube.com/watch?v=YlVDGmjz7eM)

No one coming into a technology management class at Northwestern University expected to see a comedy video their first day and I would always follow up after their jaws dropped with my comment, *"We will be focusing on the Dangerfield approach to network infrastructure. In other words, we will focus on the real world, and not Fantasy Land."*

With three minutes of video, I captured their attention so fast and they knew this course was going to be different.

CARLINI-ISM: *"Don't be politically correct. Be politically accurate."*

My courses were entertaining. They had to be. These were adults (all over 21) coming in after working eight hours in the day. I had to use different techniques to get and keep their attention.

We even had field trips to different places depending on the course. What better way to explain technology than to walk right through an operational center utilizing all the components discussed in the textbook? I had people from other universities coming to take a course and then transfer it back to their regular school. (Sometimes the undergraduate course would be transferred as graduate-level credit because no one was teaching concepts the way I was.)

Trying to understand corporate and municipal mission critical applications? We walked through the Chicago Mercantile Exchange (the trading floor), Walgreen's Data Center for all their Prescription Services (and the base of all their Satellite services), and the Chicago 911 Center which I was the consultant to the Mayor's office on its planning and conceptual design. We even took a tour through AT&T's 10 S. Canal building which is the only nuclear bomb-proof building in downtown Chicago. It also has four jet engines on the top floor to generate power for days if they lose electricity from the power company.

Talk about seeing real examples of how buildings, networks, and mission critical applications are supposed to be integrated and built correctly, these field trips were the talk of the student base. Reading a book was one thing. Walking through a live 911 center or a mission critical data center and seeing everything working is a totally different experience. The knowledge about applying the technology stays with you longer when you actually see it in action.

Think about the old management adage, MBWA (Management By Walking Around). You really learn a lot when you walk through something and not just read a book about it.

In the first course of telecommunications and computer networks, I preached you cannot go against the Five Laws of Networks. **(See CHART 2-3)**

Learn these laws because they lay out a good foundation when thinking about planning and designing a solid network infrastructure for any organization, building or business campus environment:

CHART 2-3: THE FIVE LAWS OF NETWORKS

1	Networks never get smaller.
2	Networks never get slower.
3	Networks never stay the same.
4	Networks never work all the time.
5	Networks (good ones) are never cheap.

1) **Networks never get smaller.** Any viable organization is going to be growing its enterprise-wide network as well adding new applications. If it is a dying company, it will be bought out or spun into bankruptcy.

2) **Networks never get slower.** Any viable organization is always looking at increasing speeds for new applications. You never see any company announcing that they are going to slow down their network. That's illogical. Speed is critical.

3) **Networks never stay the same.** Any viable organization is constantly changing its network topology. From adding people to new

locations, the network is always being changed. The organization could be buying another organization and adapting their network to absorb the network of the acquired company.

4) Networks never work all the time. No matter what company, including ALL the network carriers, no network has 100% reliability. We strive for 100%, but in reality there is some downtime. The rule-of-thumb downtime for central office processors was two hours for every forty years.

(That is a very high standard, but there still was allowance for outages. Don't you wish your PCs/laptops, Smartphones, and Tablets had that type of downtime goal?)

5) Networks (good ones) are never cheap. You have to spend money to make money. If you have mission critical applications, everything has to be fully redundant in your network. That is NOT the cheapest solution, but you have to worry about reliability and resiliency. There is a big price tag which goes along with having redundant resources.

People wanted to learn the right way and I never had any problems with attendance. Students always wanted to come see what was in the next class and see working mission critical networks serving different enterprise applications in various industries.

CARLINI-ISM: *"Courses aren't boring, people are. A good teacher can generate excitement about any topic."*

All that background and experience has helped in understanding the evolution of real-time systems, the need to focus on quality in software engineering, understanding emerging communications technologies, the convergence of several industries, and the public switched telephone network (PSTN) infrastructure as well as the revolution in the internet and wireless handheld devices (Smartphones and Tablets) that we employ today.

What I have seen in the last several years is there needs to be a shift in higher-level education. A multi-disciplinary focus must replace the single-focus expertise which makes up traditional commercial real estate development teams. This concept was introduced back in the late 1980s, yet many did not see the importance.

Today, redefining team dynamics and incorporating a broader focus as it relates to planning, designing and operating commercial real estate is a must. In addition, those involved in economic development must also understand a broader perspective which includes infrastructure and the application of various technologies (computers, networks, Smartphones). Traditional business advisors (lawyers and accountants) must be augmented with technologists when it comes to understanding all perspectives of a commercial or municipal development.

Both sales and technical people in the areas of technology need to understand where we came from and where the converged real estate/infrastructure/technology market for intelligent amenities is headed. Real estate, architectural, technology, and infrastructure professionals need a solid background to support this industry and all of its affiliated technologies which have now grown and diversified to cloud computing, shared resources and new edge technology like Smartphones and Tablets.

Before, traditional business advisors to executives making decisions on real estate and new facilities were lawyers and accountants. Today, they must be augmented with technologists when it comes to understanding all perspectives of a commercial or municipal development. **(See CHART 2-4)**

CARLINI-ISM: *"The use of traditional business advisors - lawyers and accountants - is not enough for assessing deals involving technology. They must be augmented with technologists."*

CHART 2-4: TRADITIONAL BUSINESS ADVISORS MUST BE AUGMENTED

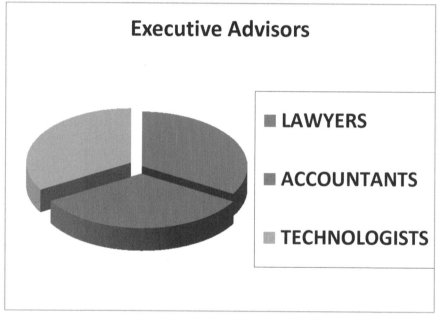

Today, we need to have access to a myriad of users across multiple networks. It has been said, *"Information needs to be at our fingertips."* That is not enough. Besides accessible information needing to be at our fingertips, access to the processing power and real-time applications utilizing that information need to be at our fingertips as well.

This is very evident in the area of Smartphones. The market leaders like Apple and Samsung have come to understand Smartphones and Tablets have become the new edge technology for people to use while doing work, doing personal business, shopping, watching videos, or just wanting to connect in some form of social media.

For almost a decade, there has been a clear consolidation as to the type of Operating Systems used in Smartphones. The two predominant ones today are Android and iOS in the United States.

According to ComScore in a 2014 study, Smartphone users in the U.S. increased 24% between 2013 and 2014, reaching 156 million users and Tablets grew 57% in the U.S. in 2013 to 82 million owners. This is a huge market shift of how people access the Internet and that impacts the need for real estate to reflect how these new devices will be handled within the buildings, campuses and business parks not only in the United States, but globally as well. The graph also shows the consolidation of Smartphone operating systems. **(See CHART 2-5)**

CHART 2-5: CONSOLIDATION OF SMARTPHONE OPERATING SYSTEMS

U.S. Smartphone Market Share by Operating System (OS)

comScore MobiLens, U.S., Age 13+, December 2005 - December 2013

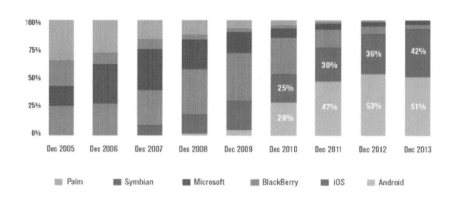

ACCESS BETTER BE REDUNDANT

Business leaders need to realize traditional rules-of-thumb for real estate and connectivity are obsolete.

The idea of a single connection to the telephone company's central office that serves the building literally goes back to the horse-and-buggy days. It is obsolete.

Today, we need to have two separate routes to separate central offices to provide redundancy and resiliency to the access of the network. Businesses need to have back-up to all mission critical applications. Core business applications are growing and there is one mission critical app for every three applications today. In a short time, that ratio should increase to one out of two applications being mission critical.

You need to take into consideration new network, as well as power, requirements for mission critical applications.

CARLINI-ISM: *"Intelligent amenities are more important than traditional amenities. Traditional amenities are a given in commercial real estate – intelligent amenities are not."*

Most buildings need to undergo a transformation as to how they connect into the fabric of network infrastructure within the region. A single connection equates to a potential single-point-of-failure. Many building owners need to assess whether or not they want to incur the expense of adding a second route for communications into the building.

CARLINI-ISM: *"A single connection equates to a potential single-point-of-failure. That being said, over 95% of buildings are then not ready for tenants who have mission critical applications which need network access."*

CHAPTER REVIEW

CARLINI-ISMS

"Whoever has the best-trained workforce, is the toughest competitor."

"The broader the perspective, the better the problem-solver, especially in applying technology to business."

"Eliminate any single-point-of-failure on any mission critical network and/or application."

"Always try to explain technical terms in everyday terms and you will get a lot more positive response as well as clear understanding from those who do not have a technology background."

"Structuring the unstructured is a large step in the planning and design of any mission critical application."

"Quality is the bedrock of all successful endeavors."

"Don't be politically correct. Be politically accurate."

"Courses aren't boring, people are. A good teacher can generate excitement about any topic."

"The use of traditional business advisors - lawyers and accountants - is not enough for assessing deals involving technology. They must be augmented with technologists."

"Intelligent amenities are more important than traditional amenities. Traditional amenities are a given in commercial real estate – intelligent amenities are not."

"A single connection equates to a potential single-point-of-failure. That being said, over 95% of buildings are then not ready for tenants who have mission critical applications which need network access."

[3]

TRANSFORMING ORGANIZATIONAL TITANICS INTO STARSHIPS

"Leading-edge organizations do not maintain their position using trailing-edge technologies." - **JAMES CARLINI**, 1984

We are past the industrial age. We are past the information age. We are in the middle of the "mobile internet" age which means we have to be able to conduct business anywhere, at any time, and with anybody. Yet in many instances, we still teach from an industrial age standpoint and it is not only in our public schools. Most universities are teaching real estate from a 1950s perspective at best, and a 1850s perspective, at worst. We need to improve our teaching methods in order to convey 21st century topics and converging disciplines like real estate, infrastructure, and technology.

Right around the corner will be age of the Internet of Things (IoT) and because of the growth in wireless devices between now and the year 2020, we will need to re-think many concepts about network infrastructure and how we integrate it into buildings, business parks and other venues.

Industrial-age solutions need to be abandoned. Traditional approaches to businesses as well as public infrastructure do not work anymore and should not be adhered to within a global economy. What is the local and regional impact if this is not followed?

We need to move higher education into the 21st century as well as the whole public school system. Today, some would say public schools are an anachronism.

CARLINI-ISM: *"Creating new conceptual frameworks for infrastructure means tearing down obsolete education and curricula."*

The public school systems were built to assimilate an agrarian workforce into industrial age jobs. We are well beyond that as a workforce as well as a nation competing in a global economy. Educational institutions need to update their curricula. From public school systems to university-level courses, industrial-age solutions need to be abandoned.

Industrial-age leadership also needs to be abandoned. Traditional management approaches to businesses as well as status quo infrastructure do not work anymore and should not be adhered to when building new buildings and business campuses for supporting businesses needing to compete within the global economy.

CARLINI-ISM: *"Building for the future means advancing from the past."*

What is the impact if this is not followed? We will continue in a downward spiral of the Vortex of Declining Mediocrity (VDM).

Something as simple as having a long-running, rule-of-thumb for buildings serviced by the phone company, *"one connection to one network carrier"* has to be changed. With more organizations requiring and establishing mission critical applications, "a single line, and single Network Carrier link" is the equivalent to a horse-and-buggy solution back in the industrial age. Revised education across many disciplines, from Architecture to Network Engineering, is needed in this area to move forward from so many obsolete rules-of-thumb.

THE THREE R'S

Let's start at the beginning. The Three R's which are still taught in many public schools are not *"reading, writing and arithmetic."* The Three R's that are still being drilled into our students (and future workforce) are the skills needed for industrial age jobs.

(SEE CHART 3-1)

CHART 3-1: INDUSTRIAL AGE SKILL SETS

SKILL SET	REASON
ROTE	In factory jobs, there were some job functions that had to be memorized by the worker.
REPETITION	Many jobs had repetitive tasks associates with them. Especially, assembly line jobs.
ROUTINE	Jobs on an assembly line were compartmentalized and once you learned your three to five job functions, it became a very routine job for the eight-hour shift.

These skill sets may have been a good foundation for developing the workforce for industrial age jobs, but not the jobs in today's market. The Three R's add up to regimentation and this comes up short for today's necessary skill sets.

We need to train people to assimilate into the post-information age economy and that includes those in higher education. We need to re-evaluate what is important to teach and we must weed out the deadwood teachers who cling on with poor skills protected by tenure.

CARLINI-ISM: *"Those who can, do; Those who can't? Teach.' This has to be replaced with – 'Those who can – must teach,' in order to promote a pragmatic, realistic perspective within the upcoming workforce."*

The skill sets needed for today's jobs goes beyond what is being taught today. The new skill sets needed to compete in today and tomorrow's global economy are fact-based. **(SEE CHART 3-2)**

CHART 3-2: FACT-BASED SKILL SETS

SKILL SET	REASON
FLEXIBILITY	Learning a couple of skills today does not set you up for a lifelong career. Lifelong learning and learning how to be flexible is critical.
ADAPTABILITY	Jobs are not "routine" anymore. Things change and the worker needs to adapt to sometimes constantly changing conditions.
CREATIVITY	How do you attack a problem? Solutions evolve as challenges change. Creative people are needed for innovation as well as defining alternatives for dynamically changing environments.
TECHNOLOGY	Every industry has been touched by computerization. Computer skills have become "basic skills" you must have in order to be viable in the workforce.

These new "**FACT**"-based skill sets must also find their way into real estate planning, marketing strategies, property management and regional planning for economic development.

The Three R's taught as skill sets of industrial-age education (Rote, Repetition and Routine) must be updated to include the skill sets of today (Flexibility, Adaptability, Creativity and Technology skills). After twelve years of being bombarded by "The Three R's", students are ready for the industrial-age jobs which required the *"Regimentation"* of those skills in their daily routine.

Unfortunately with today's high school graduates, there is a disconnect. The jobs of today, as well as tomorrow, need to be filled with people who possess a whole new set of flexible, adaptable, and creative skills using technologies in order to be successful.

Industrial-age solutions need to be abandoned. Traditional approaches to businesses as well as public infrastructure do not work anymore and should not be adhered to within a global economy. This is a hidden obstacle still present in many companies today. The "industrial-age framework" of "The Three R's" instilled in people's minds from public education is actually hindering them in trying to grasp and overcome today's complex problems in many diverse industries.

Today's and tomorrow's problems are not routine. They need to be approached with flexibility by people who have creative and adaptive problem-solving skills. Plus, the answer is probably going to be technology-based or tied to some technology-driven solution, no matter what industry you are looking at.

EDUCATION NEEDED TODAY

Today, education is important not only for those aspiring to get into this combined high-tech arena of real estate, infrastructure, and technology, but for those who want to build upon it to develop and grow businesses.

In the convergence of real estate, infrastructure, and technology, we are trying to develop a better and more concise understanding of their strategic intersections as well as their tactical underpinnings. We need to get some common grasp of the divergent elements and understand how they now interact with one another.

What was once considered diverse industries, now have common ties, technologies, and foundations which should be noted as well as understood in real estate development projects as well as existing commercial and industrial space.

This integrated concept has been percolating over the last two decades as consulting and advising work on various projects have provided a rich insight on what is important, what is gaining traction, and what is obsolete. What is needed today is a framework to discuss how everything fits together.

The concept and framework of the *"Platform for Commerce"* was first introduced and presented in my keynote speech at the broadband properties Summit '08 in Dallas, Texas in April of 2008 **(See CHART 3-3)**. The common thread across each layer of infrastructure is the transportation of goods and services on the expansion of trade routes across land, sea, air and finally electronically.

It was later formalized into a white paper, **"Intelligent Infrastructure: Securing Regional Sustainability"**, written and submitted to the U.S. Department of Homeland Security. It was selected for presentation at their Conference on Aging Infrastructure at Columbia University in New York in 2009. It has been further refined to become the recognized global *"Platform for Commerce"* in Chapter Four.

This framework depicting infrastructure is critical because it provides a more universal look at what encompasses the infrastructure for today's global economy rather than what many people hold as a partial and traditional definition of infrastructure (roads, bridges, railroads, and maybe the power grid). Economic development decisions affecting regional sustainability must include all facets of infrastructure in order for them to be viable.

CARLINI-ISM: *"Leading-edge countries do not maintain their position using trailing-edge infrastructures."*

Back in the mid-1980s, the concept of intelligent buildings emerged. This was the beginning of corporate organizations looking at commercial real estate differently and becoming more discriminating on what location would be selected to establish a facility.

At that point, many corporate entities were still moving from the industrial age to the information age in the way they managed their business and enterprise. They were becoming more communications and information technology systems-dependent, no matter what industry they were in. Computerization of a business meant major changes in the workplace environment and the facilities a building needed in order to support those new types of tenants.

CHART 3-3: PLATFORM FOR COMMERCE

LAYER	LEVEL	DOMINANT INITIAL DRIVER OF IMPLEMENTATION IMPORTANCE
SPACE (FUTURE) (INTERPLANE-TARY)	8	JUST BEGINNING TO BE BUILT (Space shuttles, station, satellite networks) (Future – mid-21st Century, 22nd Century? US, RUSSIA, JAPAN, CHINA?)
BROADBAND CONNECTIVITY NETWORK	7	CHINA, JAPAN, KOREA, NETHERLANDS, US (beginning 21st Century)
AIRPORTS	6	UNITED STATES (mid-20th Century)
POWER (GRIDS, NUCLEAR POWER)	5	UNITED STATES (beginning/ mid-20th Century)
TELEPHONE NETWORK (VOICE ONLY)	4	UNITED STATES (beginning/ mid-20th Century)
RAILROADS	3	UNITED STATES (mid-1800s)
ROADS/BRIDGES	2	ROMAN EMPIRE (500BC- 476AD)
PORTS/ DOCKS/ WATER	1	PHOENICIANS (1200BC-900BC) EGYPTIANS (3000BC-1400BC)

Source: JAMES CARLINI, KEYNOTE SPEECH, 2008, 2012 - All Rights Reserved

One of the new ideas for office buildings was the concept of Shared Tenant Services (or STS) especially in the area of telecommunications.

A private branch exchange (PBX) was an On-Premise phone system which could be segmented to serve different customers within the building, therefore eliminating the need for them to go out and buy their own phone system. A shared PBX among the building's tenants was thought to be a new profit center for the building as well as a desirable "intelligent amenity" to offer prospective tenants to differentiate the building from other properties. Subscribing to a shared phone system would give tenants access to advanced features and capabilities they may not have been able to afford in a stand-alone environment.

There was definitely room for profit in long-distance services at that time. The main focus for the building owner or property manager was to share certain communication costs across several tenants and reap some economies of scale as well as offer sophisticated services the individual tenants could not have otherwise afforded on a standalone basis. Shared phone services and shared office automation services (shared servers) were all part of the package and in a way, were the "Cloud Computing" concept of the day. There also was some focus on developing some creative ideas to expand these core services and provide some differentiation to the building's offered amenities.

Besides new communications companies, major old-line HVAC companies like Honeywell and Johnson Controls got into the shared tenant services business as a way to gain more control over their existing customer base by offering technology-based products and services beyond their traditional heating, ventilation, and cooling (HVAC) product lines. The emergence of computer-controlled Energy Management Systems (EMS) became a new product as part of the products and services sold in this "intelligent building services" market.

Other companies got into this area as well, all looking for a new way to package products and services that were focused on providing new technologies to those tenants and companies who may not have been able to afford them if they had to pay for each on a stand-alone basis. The era of "intelligent buildings" was born.

Did all real estate's executives embrace these changes at that time? No, but there was enough movement and adoption which really made a mark on the way commercial real estate was developed, managed and compared. Some walked away with new ideas and an appreciation of adding these intelligent amenities onto their buildings to attract a higher-caliber tenant, while others remained stuck in the past dependent on simplistic strategies that didn't amount to anything more than cutting back lease prices when the competition cut theirs. Not a smart strategy in any economic time. As one real estate executive pointed out, *"If you live by price, you die by price in this industry."*

The idea emerging from all this new activity was that a multi-disciplinary strategic focus must replace the single-focus expertise when integrating these capabilities within a building or building campus (industrial park). System integrators had to develop broader skills in order to understand how all of the components fit. Organizations had to re-structure their departments and recruit new types of talent.

CARLINI-ISM: *"Organizational structures become obsolete just like the technology that they manage. Management structures must be reviewed and replaced just like systems, software and technology."*

It isn't good enough to be a property manager, an electrical engineer, a mechanical engineer or a network engineer today.

In order to understand how to apply the complex strategies of today, you have to understand how concepts from different disciplines are inter-related and merged together in order to create the support systems and intelligent infrastructure within a building as well as a business park/campus.

The same holds true for real estate and property management executives. They have to understand how these capabilities and "intelligent amenities" augment their buildings' traditional amenities and their overall marketability. Intelligent amenities like broadband connectivity, shared computer resources and other new shared tenant services (like sharing a diesel generator for back-up of mission critical systems) have to be understood on how they would affect the design and cost of the building as well as the marketing and leasing of the building to potential tenants.

Unfortunately, the real estate market really dropped around 1990 and many of these new endeavors and the implementation of intelligent building concepts as well as the high-tech organizations and their management structures fizzled out. Many of the concepts however, were still very relevant and became more relevant as time went on with companies becoming more dependent on high-speed connectivity and mission critical applications for doing business in a global economy.

Today, redefining team dynamics and incorporating a broader focus as it relates to planning, designing, operating, and differentiating commercial real estate is a must. In addition, those involved in economic development must also understand a broader perspective and the impact of technology and broadband connectivity on real estate.

Today, world-class broadband connectivity and wireless connectivity are becoming critical requirements in many new Intelligent Business Campuses (IBCs) and other next-generation real estate venues aimed at enticing sophisticated corporations, their corporate facilities and their technology-savvy personnel. It is definitely a distinct competitive advantage now for multi-venue environments which include retail, entertainment, and convention centers, but it could turn into a competitive necessity very shortly as more people opt to go with Smartphones and Tablets as their new edge technology.

A higher level of infrastructure is being desired by tenants. Corporate users of technology are not buying off on just "traditional amenities" anymore. (Today, you cannot get away from this fact.) They expect the intelligent amenities along with traditional amenities as part of the entire package. This means integrated systems combining functional capabilities and economies of maintenance for the building must be implemented and maintained by the owner/property manager.

Traditional amenities like parking, elevators, security and air conditioning were taken for granted that they would be part of what a building offered. The new "intelligent" amenities focused more on supporting the organization's communication and information needs, especially when they have mission critical applications.

CARLINI-ISM: *"New skills are needed to solve information age endeavors. Do not rely on the 'traditional methods' of management."*

Comparing intelligent amenities like information technology services, network communications, power and building automation systems to assess what property would best suit an organization looking for a new location began at this point.

Comparing buildings for their offerings and availability of intelligent amenities began in 1985. I derived a Building IQ test at that time for a major client, JMB Property Management, who was based in Chicago. The test was to insure they were getting their money's worth on a $20,000,000 retrofit of the Seattle Seafirst Building in Seattle, Washington.

They had contracted with Honeywell to improve the HVAC systems within the building as well as provide new building controls and network infrastructure within the building.

JMB wanted to know if they were getting their money's worth, plain and simple. They contracted me and I developed a comprehensive test that questioned what each surrounding competitor's building had to offer in three main categories: Information Technologies, Communications, and Building Automation.

Many within the real estate business as well as supporting industries were new to this approach and were caught off-guard when asked questions about their building's intelligent amenities. They were still focused on traditional amenities and traditional marketing strategies to sell space in buildings. (**See CHART 3-4**: Comparison of Traditional and Intelligent Amenities)

At the time of the comparisons, JMB was initially worried about making sure they got their money's worth on the technology purchased for the retrofit project they initiated, but as soon as they found out their building was technologically superior in its systems and infrastructure, they wanted to publicize the test results and show the downtown Seattle market that their building was significantly better than the ones around it. Offering technology became a marketing differentiator. *"How intelligent was the building's systems?"* became an important question from potential tenants comparing leases and space in downtown markets.

CHART 3-4: COMPARISON OF TRADITIONAL AND INTELLIGENT AMENITIES

TRADITIONAL AMENITIES	INTELLIGENT AMENITIES
Parking Spaces	Parking, electric vehicle battery hook-ups
Elevators	WiFi/DAS networks
Basic communications (single carrier, single feed (single point-of-presence))	High-speed connectivity (multi-gigabit speeds), multiple network carriers, dual feeds (dual points-of-presence), wireless
Access to Highways and Transportation	Access to multiple network carriers
Power to Campus (adequate single source)	Power (access to multiple source, on different grids)
Security (Locks, gates)	Security (electronic card access, IPTV surveillance)
Space (little or no adaptability to newer technologies)	Space (adaptable and conducive to technology change)
Power to individual building (adequate)	Power to individual building (adaptive, redundant, and alternative source)
HVAC	Shared power back-up generation, more sensors and controls for more control points for energy savings

Some of the executives did not see the interrelationships formed by the mix of amenities that supported a building or campus of buildings. These interrelationships are more noticeable today and the need to implement the right mix of intelligent amenities is critical today in order to attract and maintain corporate tenants.

Redefining team dynamics as they relate to planning, designing and operating commercial real estate is also a must today in order to understand and effectively manage those interrelationships.

Before intelligent buildings, traditional business advisors to executives making decisions on new real estate facilities were lawyers and accountants. Today, technologists must also be included as strategic business advisors on major deals, acquisitions and developments - not just lawyers and accountants. This is not only true for real estate endeavors, but for any large technology investment in any organization's technology infrastructure.

Another major example of needing a technologist to "bless the deal" was when the Williams Companies decided to sell access to their obsolete pipeline to a network carrier so they could pull fiber optic cable throughout the pipeline's conduit throughout 17 states. The accountants looked at the numbers on the deal and the lawyers looked at the contract making sure "the I's were dotted and the T's were crossed", but they needed a technologist to approve the value that they were getting for the use of the obsolete pipeline and its right-of-way.

At about the same time of the JMB project in 1986, the Intelligent Building Institute (IBI) was launched in Washington, DC as a non-for-profit organization where various companies could bring in their perspectives to help shape this new and emerging industry within the Real Estate industry.

I was a member and the Chairman of the Definitions Committee of that organization from 1986-1988. One of the issues I saw was the importance of the development and adoption of common definitions across different industries. We needed to get everyone starting on the same page. That was a difficult task then, and still is today.

Understanding what other members of a development team (the architect, mechanical engineer, electrical engineer, networking and systems integrator) were talking about was critical to the planning and development of an Intelligent Building.

CARLINI-ISM: *"Today, many people do not know how we got to this point in applying technology. You have to know what led up to today's configuration and applications of technologies."*

Unfortunately, when the real estate market bottomed out around 1990, many who were involved in Intelligent Buildings went on to other endeavors. Real estate executives were not interested in adding intelligent amenities to their buildings even though they would have benefitted from them. They were just concerned about survival in the near-term. We did not get to that point where everyone was on the same page when it came to applying technology to buildings.

Today, some of those concepts need to be re-introduced in order to understand how we got to where we are at today. Basic definitions, metrics, frameworks and templates must be clearly understood across the board as well as readily applied in order to maximize potential of these converged areas. We need to start everyone on the same page again in order to move forward more quickly.

Applying Technology to Business Is A Continual Process

Applying technology to an enterprise is not a "one-shot deal", it is a continual process. You need to continually look at what is happening and decide what elements of supporting technology you need to update. This will make a difference in whether or not you lead your industry or remain part of the pack chasing the industry leader. Adding key technology can give you a competitive advantage. As time goes on, that technology becomes a competitive necessity in order to compete in the industry. If you try to keep using obsolete technology, it is a competitive disadvantage. **(See CHART 3-5)**

If an organization wants to "break out of the pack", it can establish a strategic directive to add new capabilities onto its organization's infrastructure. This will give them a competitive advantage over their competitors. As time goes on, this unique competitive advantage might be matched or surpassed by competitors. If it is matched, it becomes a competitive necessity – something which is needed in order to remain viable in the industry. If it is surpassed by new technologies, it becomes a competitive disadvantage to continue maintaining this as part of the organization's infrastructure.

Social Media Tools Are Not All Tools

With the use of new social media tools like Twitter, Facebook, and LinkedIn, we are seeing new types of social communications beyond what we would consider "traditional" communication forms.

There is not enough historical data of usage and applications on most of these "tools" to say that they are all effective. We are still in the process of sorting them out as tools or toys. The overnight experts in social media claim all social media tools are critical in today's and tomorrow's markets, but that is still up for judgment.

What we can derive from all these new tools is everyone is more dependent on communication devices and the new edge technology is more focused on Smartphones and Tablets, rather than cell phones and laptop PCs. **(See CHART 3-6)**

There is definitely an evolution taking place and there needs to be a good analysis of what makes sense for enterprise-wide communications as much as individual communications as well as where they begin to merge and converge.

CARLINI-ISM: *"Fads fade fast. Learn how to distinguish between toys and tools."*

CHART 3-5: THE CONTINUAL PROCESS OF ADDING TECHNOLOGY TO THE ENTERPRISE (OVER TIME)

COMPETITIVE

ADVANTAGE

Investment has been made and now your technology exceeds your competitions'. You lead the pack.

COMPETITIVE

NECESSITY

Technology is in-place to compete in the industry but you are only "part of the pack" trying to keep pace.

COMPETITIVE

DISADVANTAGE -

You are maintaining trailing-edge technology. You are trailing the pack and are at a competitive disadvantage within your industry.

CHART 3-6 is only a snapshot of what is currently being done. After a period of time, those behaviors will change and there will be a different analysis of how tools are being used. In another two to three years, some may fall out of favor completely and new tools that are not even mentioned may evolve into new applications. Credit: comScore.

CHART 3-6 – TIME SPENT ON SOCIAL NETWORKS DESKTOP vs. MOBILE

U.S. Share of Time Spent on Social Networks Between Platforms

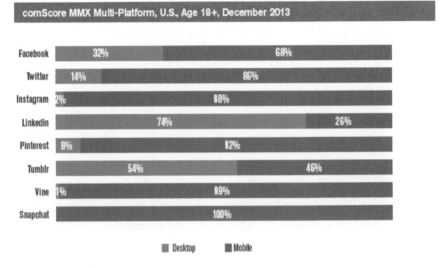

comScore MMX Multi-Platform, U.S., Age 18+, December 2013

	Desktop	Mobile
Facebook	32%	68%
Twitter	14%	86%
Instagram	2%	98%
LinkedIn	74%	26%
Pinterest	8%	92%
Tumblr	54%	46%
Vine	1%	99%
Snapchat		100%

NEW CONCEPTS IN DEVELOPING REAL ESTATE

Reviewing over 42,000,000 square feet of commercial real estate from Seattle, Los Angeles (Century City), Phoenix, and Beverly Hills to Chicago; Milwaukee; Columbus, OH; Kansas City; and Las Vegas as well as research in Asia on next-generation Intelligent Business Campuses (IBCs) and Intelligent Industrial Parks (IIPs) has given me a broad perspective on how all these elements are converging.

This book is my way of transferring that broad perspective to you so you can develop a broader perspective in a shorter period of time than it took me.

The remainder of this book will discuss new development, procurement and planning concepts along with master planning issues, infrastructure, and strategic marketing approaches in developing a *"Platform for Commerce"* which effects regional economic development.

The critical impact of layers of infrastructure will be discussed as the global *Platform for Commerce* in detail as well as the **TARGET** Map of Technology which has been used as a strategic conceptual model to get the point across that no matter what, "Technology And Revolutionary Gadgets Eventually Timeout" (**TARGET**). This model has been used as part of coursework and discussions in both undergrad and Executive Masters programs that focus on strategic planning and managing technology within an enterprise as well as for municipal government.

Chapters will discuss origins and implementations of concepts as well as the significance of understanding them from several important perspectives:

- Tenant perspective;
- Real Estate developer's perspective;
- Economic Development perspective;
- Local government (municipal) perspective; and
- Corporate and residential tenant perspective.

From the tenant perspective, there are several main issues:

- Connectivity demands
- Security
- Convergence
- Attractive costs – pricing models

- Applications
- Mission critical needs

From the real estate developer/ property owner's perspective:

- Market differentiator
- Provides a new magnet for tenants
- Competition for tenants is high
- Procurement of technology

From a municipal perspective:

- Need for economic development
- Need for sustaining future viability
- Global competitiveness
- Networks have become part of the infrastructure for commerce

We will take a quick look at the past, the present, and the future of infrastructure, technology within real estate and the impact of their convergence on trade and global commerce. We will also come to understand why, "Location, Location, Location" has become "Location, Location, Connectivity" as to what is most important in real estate today.

CARLINI-ISM: *"Professionalism is not a degree or certifications; it is a state of mind."*

CHAPTER REVIEW

CARLINI-ISMS

"Creating new conceptual frameworks for infrastructure means tearing down obsolete education and curricula."

"Building for the future means advancing from the past."

" 'Those who can, do. Those who can't? Teach.' This has to be replaced with –'Those who can – MUST teach, in order to promote a pragmatic, realistic perspective within the upcoming workforce.'"

"Leading-edge countries do not maintain their position using trailing-edge infrastructures."

"Organizational structures become obsolete just like the technology that they manage. Management structures must be reviewed and replaced just like systems, software and technology."

"New skills are needed to solve information age endeavors. Do not rely on the "traditional methods" of management."

"Today, many people do not know how we got to this point in applying technology. You have to know what led up to today's configuration and applications of technologies."

"Fads fade fast. Learn how to distinguish between toys and tools."

"Professionalism is not a degree or certifications; it is a state of mind."

[4]

LOCATION, LOCATION CONNECTIVITY

"Real estate's three most important words."

"Those who have the creative ideas, have no capital. And those who have the capital, have no creative ideas." - **ANTHONY V. CACCOMO, AVC REALTY**

Today's economic development equals broadband connectivity and broadband connectivity equals jobs. From a political standpoint – jobs equal votes. This paradigm shift needs to be recognized by those in all aspects of real estate development, economic development, infrastructure, and technology as they converge to address regional economic sustainability and global commerce.

What used to be industrial parks and business campuses which were sold and leased to businesses have become more sophisticated. There are many companies looking for more than just space.

In an interview with BUSINESS 2.0 magazine in December 2004, I proclaimed the old real estate adage of *"location, location, location are no longer the three most important words in real estate today."* In the article, I pointed out that the real estate adage for those trying to analyze where to start or expand businesses used to be *"location, location, location." now the rule-of-thumb is, "location, location, and connectivity, because of their dependence on communication-based applications that support their businesses."* This not only applies to commercial and industrial buildings, it applies to other real estate as well. **(See CHART 4-1)**

CHART 4-1: TYPES OF REAL ESTATE

TYPES OF REAL ESTATE	TRADITIONAL EXAMPLES
COMMERCIAL	Single buildings (hotels, theaters, office buildings) and business campuses, shopping centers, parking
INDUSTRIAL	Industrial parks, power plants, warehouses, tech campuses
RESIDENTIAL	Single family homes, townhouses, multi-tenant apartments/ condos
AGRICULTURAL	Farms, ranches, nurseries, timberland
SPECIAL PURPOSE	Schools, churches, other special purpose

Since 2004, people have gotten very dependent on being "connected" through the internet as well as personal communication devices and if a multi-tenant building or even a single family dwelling cannot get access to high-speed connectivity, it is falling out of favor by a growing number of people ranging from young professionals to senior citizens who all have become dependent on more and more technology applications, Smartphones, Tablets, and other internet-based social media tools. This dependence is still growing as we enter the era of the "Internet of Things" (IoT).

Today, the three most important words in choosing where your company or organization should be located in commercial real estate are "location, location, connectivity." It really applies to where you establish your residence as well when you think about all the people who work out of the house.

"Location, location, connectivity" are the three most important words because they are the new standard for significance in reviewing and selecting good locations for corporate facilities as well as residential housing. The new "three words" translate into the political criteria required for regional economic sustainability as well.

If you want to attract and maintain new corporate facilities for your region, broadband connectivity better be available within the fabric of the local infrastructure. Job creation, and all it entails, starts by insuring you create an infrastructure which is strong enough to provide a solid **Platform for Commerce** including a layer for broadband connectivity for all industries.

If property owners and commercial property management companies do not have access to high-speed broadband connectivity for their buildings, they are going to lose out on vying for the top corporate tenants of today as well as tomorrow. They will also lose their current first-tier tenants to buildings in regions which provide new "broadband connectivity." This applies to office complexes as well as industrial parks.

Today, high-speed broadband connectivity (defined as anything over 1 Gigabit of speed per second) is one of the top three criteria for corporate site-selection committees to ascertain when reviewing potential sites for their organization. The comparison between various line speeds can be seen in the speed chart comparison. **(See CHART 4-2)**

I developed that "Speed Chart Comparison" for a Chicago Tribune reporter who was looking for a way to convey what the higher gigabit speeds meant to an average user. At the time, he was interviewing me on my work with the DuPage Business Center several years ago and nothing was available to give an everyday example for a comparison.

This growing requirement applies to residential properties as well. Most young professionals want internet access as part of their amenities in multi-tenant buildings, especially those who telecommute for their jobs. Even older couples want to make sure they have broadband connectivity not only for themselves, but for their children and grandchildren as well. It will only grow in importance into the future. If your building doesn't have it, the marketability of your building will shrink along with its value.

In my own experiences as president of the condo association for a 72-unit luxury condo building on Lake Michigan in Wisconsin, adding WiFi to the common area party rooms for everyone to access was just about a stroke of genius as to upgrading the building's offering of amenities and keeping it up-to-date without re-wiring the whole building. Those who wanted access could simply go to one of the party rooms to access high-speed connectivity without having to subscribe to it for their own condo.

CHART 4-2: SPEED CHART COMPARISON OF NETWORK SPEEDS

Speed of Transmission *Downloading a 90-minute movie (1 gigabyte*)*	
Speed of Circuit (type)	Time Elapsed (rounded)
56 Kbps (dial-up)	426 hours (~ 17.7 days)
1.5 Mbps (DSL, cable, T-1)	15.91 hours
10 Mbps (wireless)	2.39 hours
1 Gbps (fiber to the curb)	8.59 seconds
10 Gbps (fiber to the house)	0.86 second

Source: James Carlini

* Gigabyte: a unit of information equal to 1 billion (actually 1,073,741,824) bytes (or 8,589,934,592 bits) or 1,024 megabytes (or 2 to the 30 power).

With over half the owners owning a second property, it fit the building demographics nicely because many were only in the building half the year. We also got the adjoining marina to participate and subsidize the monthly fees.

Talk about adding intelligent amenities to make everyone's investment a little more viable, this was an inexpensive addition which added a whole new high-speed connectivity amenity for every condo unit and boat owner.

Sometimes, something as simple as this can bring new life and add new value to a 25 to 75 year-old building. Bottom line, it gives you something else to point to as an amenity when competing within the market and comparing building amenities.

When it comes to analyzing how the infrastructure of a region or country can affect its viability to trade in the global marketplace into the future, it is simply put as seeking what I have coined as the "Foundation of Truth":

CARLINI-ISM: *"Economic development equals broadband connectivity and broadband connectivity equals jobs."*

One of the problems is most real estate executives do not want to be the first ones implementing a new concept. The attitude at the executive level at many real estate firms is: *"Change is great – you go first."* With this attitude, most real estate executives have been slow to embrace the changes in value and the new amenities to counter the loss in value.

A MAJOR SHIFT HAS HAPPENED IN COMMERCIAL SPACE

With the emphasis on communicating anytime, anywhere and with anyone about anything, real estate must modify its approach in providing amenities for serving corporate tenants within their buildings and business campuses.

In the last 30 years, intelligent amenities which provide communications have become interwoven into the fabric of the building infrastructure because more corporate mission critical applications have become more important to sustain the core business of the enterprise.

Intelligent building concepts which were developed, proven, and then somewhat discarded in the late 1980s and early 1990s re-emerged to become important again in the 21st century. One of the original concepts of Shared Tenant Services (STS), saving on long distance services is not important anymore but, the concepts of having the latest wired and wireless network technologies as part of the network and information infrastructure within the building which support the tenants' space are critical.

Intelligent amenities are important again. Also, the sequence of tenants brought into a multi-building, multi-tenant campus is critical because the order will also dictate what types of intelligent amenities are going to be desired.

Back in the mid-1980s, measuring building intelligence was one of the areas I pioneered.   This new assessment affected marketing commercial real estate properties. Comparing commercial properties to understand which one would be the best fit for a potential tenant requiring communications-based information systems goes back to 1985 and 1986. (One of my first articles about this, comparing buildings in Seattle and the concept of intelligent amenities versus traditional amenities, appeared in *COMMERCIAL RENOVATION*, back in 1986.)

At that point, strategies for marketing real estate should have changed. Land and office space should have not been the only comparables when deciding what place to put a corporate facility.

It's taken almost three decades from the first endeavor of comparing building infrastructures and their intelligent amenities for the tenant market to take hold as a major issue across a major segment of the real estate market.

The bottom line is old real estate and economic development rules-of-thumb must be replaced. There needs to be a renaissance in the way real estate is developed and leased to insure regional sustainability. Technology and intelligent amenities must also be understood and applied as they are "must have" amenities, and not just a "hoped for" amenities, in commercial real estate. This is being proven today out in the international markets as well. As more mobile workers want access to work as well as their play from their residential units, residential real estate must also be re-evaluated and designed differently.

CARLINI-ISM: *"Municipalities are holding on to some white elephants, they just don't know it – yet."*

NEXT-GENERATION BUSINESS CAMPUSES

What is put in first, dictates what is put in next. Creating a theme for the next-generation business campus, the Intelligent Business Campus (IBC), is important. Building a business campus with a common goal actually will strengthen the property, increase its lease revenues, and insure its long-term viability. The commonality of capabilities to provide services for an all-medical corporate park or an all-legal corporate park can really make a difference in attracting and maintaining quality tenants.

Master Planning for these new developments has also taken on some new insights. The acceleration of detailing real requirements of power and telecom upfront should be added to the process of Master Planning instead of waiting until tenants are signed up. Power and network connectivity used to be afterthoughts in traditional business parks and campuses. Today, those amenities need to have been thought out and put into place as part of the supporting infrastructure which is reviewed upfront.

The new rules for real estate master planning include:

1) Telecom and power are driving factors - and must now be viewed as part of the upfront master planning process.

2) Network infrastructures must be done upfront, instead of as an afterthought after tenants have moved in.

3) Power infrastructures must be done upfront instead of as an afterthought.

4) The acceleration of power and telecom to the forefront of real requirements includes looking at adding redundant feeds rather than the traditional single feed to the central office and the substation (power grid).

5) Any next-generation business park has to have cyber-resiliency with workplace recovery capabilities.

6) Streamlined development process for tenants - must coordinate with local municipalities/ building permits/ utilities/ impact fees in order to have a "fast-track" offering to potential tenants.

7) Additional support structures (for example: a rapid response team for technology support – "one-stop shopping" morphed into "one-stop supporting" for tenants.)

8) Higher bandwidth which allows for higher and broader use of software (think more apps in Smartphones and Tablets being activated that are supported by a capacity-intensive network)

These types of next-generation parks facilitate more cohesion for the entire campus and supports adding velocity to the acceleration of technology transference from research to retail. Moving a product from concept to prototype to manufacturing and distribution can be done at one Intelligent Business Campus (IBC). This approach is being implemented around the world in several countries.

CARLINI-ISM: *"If you are looking for a building or park to support your enterprise's mission critical applications, over 95% of buildings are technologically obsolete. They have single vendor connectivity which in turn creates a single-point-of-failure."*

PARKS OF THE PAST

Older business and industrial parks did not have these strong commitments to diverse power sources and redundant network connectivity when they were initially planned. These are the ones you see all the "for sale/for lease" signs in front of the property. They are all lacking tenants and those property managers who filled them up with 20th century strategies cannot seem to figure out why they cannot lock in a new tenant today in the 21st century. "Space available" is not a sign that screams we're making money here on this property. It also screams what is for sale is a commodity – space.

Can the properties be retrofitted to compete in this next-generation approach? Yes, if the right investments are made. If not, they are destined to attract lesser tenants and generate lesser revenues. In effect, they have become technologically obsolete.

The next conversion for them may be to turn them into storage facilities where buildings are divided up into little cubicles to rent out.

Having an array of intelligent amenities offered by a building or business campus is like having more tools in a tool belt. The more building owners and property managers can offer tenants, the more tenants they can service successfully. **(See CHART 4-3)**

The property owners and property managers must realize that it is an ever-changing commitment and should look at this as being part of the overall property management of the facility. *"How smart a building do you want?"* Has to be as easy a question to ask from the leasing agent's perspective as, *"How much space do you want to lease?"*

CHART 4-3: TRADITIONAL VS. INTELLIGENT BUILDING STRATEGIES

TRADITIONAL BUILDING STRATEGY	INTELLIGENT BUILDING STRATEGY
Compete with Local Buildings, local markets	Compete with local and far away buildings, markets
In competitive times, cut prices per square foot	In competitive times, add value, add amenities
In tight times, cut prices per square foot	In tight times, sell the value of the amenities
Be focused just on real estate	Be focused on critical success factors that impact real estate
Building clusters – no focus, any tenant,	Building clusters – focus on theme, industry, targeted tenant
Master-planning does not include power and network infrastructure planning	Master-planning includes power and network infrastructure planning upfront
Single connections to both power and network communications infrastructures. (Horse-and-buggy days)	Redundant connectivity for communications and dual sources for (electrical) providers

Certain technology is only a temporary advantage into the next generation and then it has to be upgraded and/or replaced.

Over time, a competitive advantage turns into a competitive necessity and if there is no more investment into the facility those elements which need updating can degrade into competitive disadvantages.

CARLINI-ISM: *"If it takes a village to raise a child, then it takes a region to raise economic viability for all."*

What has become a fact over the years is that organizational management) structures become obsolete as well, just like the technology they manage.

These management structures must be reviewed and replaced, just like hardware and software. Those executives who have fresh ideas and are concerned about staying dynamic, must populate the organization.

Just "selling land" for office buildings, is an obsolete marketing strategy. Land is a commodity. Real estate developers, as well as government economic development committees, need to understand that a facility with supportive infrastructure is not a commodity. It is more unique. It has value-added infrastructure and intelligent amenities connected to it. Intelligent amenities are now features of quality – differentiating factors which further refine the selection process.

If the definition of mission critical is redundant facilities, every building which only has a single connection to a single network carrier and a single source for power from one electrical utility is virtually obsolete from a technology perspective. That sums it up for about 95+% of the commercial buildings in the United States."

CREATIVITY IS KEY

In one real estate project in Chicago, the tenant was a 24-hour a day call center for a major corporation. They required redundancy in both network connectivity as well as power sources and the building owner was trying to figure out how he could make his building more conducive to their "mission critical" needs. One of his solutions was to offer putting in a diesel generator into the basement of the building to have an alternative power supply for their call center. It was not a bad idea and we thought this could satisfy this major tenant.

The more we looked at this tenant issue, the more creative we became. The building was right next to the Chicago Transit Authority (CTA) lines (Mass Transit) and we knew that was powered from a totally different power grid than the building itself. We approached the CTA with the idea of connecting into their power grid for emergency use only and they agreed to the supplying of power on an emergency basis only. It was a "win-win" for the property owner. Now he could point out to a totally different source of power for tenants who needed redundancy in their power to support for their mission critical applications and he retained a first-class tenant.

This type of creative approach needs to be understood by those in real estate today as more tenants have mission critical applications and require redundancy of power as well as communications. Having redundant sources makes a building or campus more valuable and attracts a much higher caliber of tenant.

With next-generation real estate, the whole is greater than the sum of the parts. Having intelligent amenities available, having them have redundancies built in, and having tenants who require these extra capabilities creates a more profitable real estate, but more importantly, it adds to the regional economic viability. This applies to single buildings as well as buildings clustered in a park or campus environment.

The park amenities magnify synergies that create the final environment. As one property development executive put it as he commented about the DuPage National Technology Park, *"the whole is greater than the sum of the park."*

ASKING TEN QUESTIONS

Looking for a multi-venue project approach? We were looking for a solution for a stadium as well as a convention center and retail center in one municipality. We asked the network vendor's perspective on several issues:

1. Should DAS (Distributed Antennae Systems) be combined with WiFi systems in order to provide more coverage for a stadium? Is this a good design approach or is it flawed? I have heard there are issues. Please be specific as to the flaws.

2. What speeds were you designing for at the other stadiums/ venues? How did your solution address video capacities? Or did it?

3. What is the overall systems guarantee? In security, reliability, performance, and longevity as the deployment of 5G Networks loom on the horizon (2020).

4. What were the speeds you were focused on to the end-user?

5.   What capacities (maximums) were you focused on to provide service to?

6.   Do the main core components (like the equipment room and/or head-end unit) have redundant power feeds? Redundant network carrier feeds? Why or why not?

7.   How many simultaneous users can be on the system before it starts to degrade and the QOS (quality of service) deteriorates? This would depend on type of user call- voice, data, or video. How many voice calls versus how many video downloads? (Is there a rule-of-thumb when it comes to comparing voice traffic to video traffic? (i.e. For every 20 voice calls, one video call can be handled or other rule-of-thumb?)

8.   Thresholds – this ties in with Question 7. What are the thresholds for types of traffic (Voice, Data, and Video) before the system performance is degraded? Or totally maxed out?

9.   What were some of the issues (if any) with positioning antenna and insuring there was enough antennae for adequate (or superior) coverage? What is the numerical difference of hardware for an "adequate" system versus a superior system?

10.   Any other issues that you think are important in comparing systems would be appreciated. (Especially when it comes to being prepared to transform the existing infrastructure into a 5G network deployment)

These ten questions are a good starting point, but be sure to come up with your own set of questions which focus on your own situation. These are the types of questions needed to be asked as well as fully answered in today's and tomorrow's technology-driven real estate market. This was not the case twenty, or even ten, years ago.

Also remember this. There is no such thing as a *"Universal Solution"* or magic bullet in any technology and in any market.

CARLINI-ISM: *"There is no such thing as a universal solution. What works down the street may not work for your environment."*

There are four stages of technology assimilation. **(See CHART 4-4)** Each stage lags behind the earlier stage.

The first stage, **RESEARCH**, is "experimentation and development." Many people are going out in different directions and some of the ideas will get adopted and some will not.

Lagging behind is the second stage, **ACCEPTANCE**, or "adoption." It is a subset of all the technology being experimented with. Not every technology in the first stage (research) reaches into the second stage (acceptance). Some ideas get discarded along the way of getting them to a pilot program or actual implementation. Think of a funnel and the ideas being poured into the top, but the actual prototypes and products coming out of the bottom are only a small subset of all the ideas that were initially worked on.

CHART 4-4: THE STAGES OF ASSIMILATION OF TECHNOLOGY INTO SOCIETY

Stage 1	**RESEARCH**
Stage 2	**ACCEPTANCE**
Stage 3	**REGULATION**
Stage 4	**ENFORCEMENT**

Lagging behind **"ACCEPTANCE"** is the third stage: the **"REGULATION"** of the technology. This is where there needs to be some regulatory issues put into a framework to "manage the technology" and define its marketing territory.

The fourth stage is the **"ENFORCEMENT"** of the regulation on the technology. Again, this enforcement usually lags behind the regulation(s).

CARLINI-ISM: *"Think of technology assimilation as being RARE. The four stages are Research, Acceptance, Regulation, and Enforcement of the regulation."*

CHAPTER REVIEW

CARLINI-ISMS

"Economic development equals broadband connectivity and broadband connectivity equals jobs."

"Municipalities are holding on to some white elephants, they just don't know it – yet."

"If you are looking for a building or park to support your enterprise's Mission critical applications, over 95% of buildings are technologically obsolete. They have single vendor connectivity which in turn creates a single-point-of-failure."

"If it takes a village to raise a child, then it takes a region to raise economic viability for all."

"There is no such thing as a universal solution. What works down the street may not work for your environment."

"Think of technology assimilation as being RARE. The four stages are Research, Acceptance, Regulation, and Enforcement of the regulation."

[5]

THE FRAMEWORK OF CRITICAL INFRASTRUCTURE

"If it takes a village to raise a child, then it takes a region to raise economic viability for all." – **JAMES CARLINI**

Infrastructure has been the *"Platform for Commerce"* and economic development for over 5,000 years. Anyone in government concerned with regional economic development, creating jobs and expanding trade routes (both real and virtual) must understand the history, structure, and current and future applications of the *"Platform for Commerce"* framework as trade routes have become electronic and the way people conduct business has evolved.

If jobs creation is a real strategic initiative and not just a hollow campaign slogan, understanding this chapter is a must for every politician and regional economic development commission member who holds office from the large metropolitan areas down to the small rural towns where having some good-paying jobs locally can make the difference in being a viable community in the 21st century or a ghost town.

It is also important to anyone who touches regional economic development from any perspective including real estate developers, property management firms, corporate site selection committees, utilities, corporations, and everyone else concerned about planning, designing, building, and maintaining commercial space that can attract and maintain corporate facilities (which in turn create jobs and careers).

How Do We Expand Our Trade Routes And Increase Local Commerce?

This has been a question which has been asked for over five Millennia by countries and civilizations across the globe. Even today, it still is an important question for the economic survival in a global marketplace. Every village and town should be asking this – not just the top 20 or 50 cities.

Throughout the ages, trade routes were considered important to supporting the regional economic development and sustainability of every civilization. From the Egyptian, Phoenician and Roman empires to the Chinese, Europeans and the United States, trading goods and services was paramount to civilizations to thrive and survive. The economic goal of every civilization was, and still is, to increase commerce and develop new trade routes.

In the middle of last Millennium (1500s), expansion of their markets were the goals of European countries and their explorers following Marco Polo, who went overland to the Far East to establish trade. In the Renaissance, European explorers like Christopher Columbus, Vasco da Gama and others wanted to find more efficient sea routes to the Far East to maximize the trade and commerce taking place with China.

In the 1400s and 1500s, there was the great race to find better water routes to the Far East. Some explorers looked around Cape Horn in Africa and others following Columbus wanted to find that western sea route to take them directly to China and Japan.

To put it in a historical perspective, **CHART 5-1** gives an overall view of the expansion of commerce and the eight layers of critical infrastructure that supported its growth and the development of new trade routes. The *Platform for Commerce* framework provides an approximate timeline of the development and growth of critical layers throughout the last five millennia.

Space (interplanetary trade) has yet to be conquered from a realistic commerce perspective, but we might as well add it into the overall definition of the framework because it will certainly play some role in the next century (the 2100s). However for this discussion, we will focus on the broadband connectivity level.

Most people have not yet equated network infrastructure with the rest of the layers of critical infrastructure which have been recognized throughout centuries as needed for the expansion of trade and global commerce. The need to understand how to maximize those electronic trade routes is critical to maximizing the economic viability of a region as well as a country.

With any investment into the infrastructure, all layers must be reviewed and prioritized as to what is needed and what will be the residual payback, if any, compared to the initial spending.

CARLINI-ISM: *"'Building for the future' is a very hollow statement, when funding is aimed towards maintaining the past."*

This *"Platform for Commerce"* framework depicting infrastructure is critical because it provides a more universal look at what encompasses the infrastructure for today's global economy rather than what many people hold as the traditional definition of infrastructure (roads, bridges, railroads, and maybe the power grid). Economic development decisions affecting regional sustainability must include all facets of infrastructure in order for them to be viable. For the rest of the book, I will refer to it as the **Platform for Commerce** for brevity's sake.

CARLINI-ISM: *"To better see into a complex future, a clear framework is needed to structure new concepts."*

CHART 5-1: INFRASTRUCTURE:
THE PLATFORM FOR COMMERCE

LAYER	LEVEL	DOMINANT INITIAL DRIVER OF IMPLEMENTATION IMPORTANCE
SPACE (INTERPLANE-TARY) (FUTURE)	8	JUST BEGINNING TO BE BUILT (Space shuttles, space station, satellite networks) (Future: mid-21st century, 22nd century? US, RUSSIA, JAPAN, CHINA?)
BROADBAND CONNECTIVITY NETWORK (CYBERINFRA-STRUCTURE)	7B (wireless) 7A (wired)	CHINA, JAPAN, S. KOREA, NETHERLANDS, US (beginning 21st Century, IBCs, IIPs & IRECs)
AIRPORTS	6	EUROPE, UNITED STATES (mid-20th Century)
POWER (GRIDS, NUCLEAR POWER, OIL)	5B (Nuclear) 5A (everything else)	UNITED STATES (beginning/ mid-20th Century)
TELEPHONE NET-WORK (ANALOG VOICE ONLY)	4	UNITED STATES (beginning/mid-20th Century)
RAILROADS	3	UNITED STATES (mid-1800s)
ROADS/BRIDGES	2	ROMAN EMPIRE (500BC- 476AD)
PORTS/ DOCKS/ WATER	1	PHOENICIANS (1200BC-900BC) EGYPTIANS (3000BC-1400BC)

Source: JAMES CARLINI, 2009, 2013. All Rights Reserved

Infrastructure's Influence On Civilization

The Egyptians and Phoenicians were the first civilizations who had a significant regional impact on their Mediterranean neighbors. They developed the first layer of the platform for commerce by building ports using their ships to transport and deliver goods across the water thousands of years ago to expand their markets, gain economic power and spread their culture as:

"They arrived from a foreign land, bringing with them imported knowledge and skills. These they applied to their new environment, adding new cultural advances learned locally. Having excelled at seafaring in a sea-turned land, they traveled and traded widely, thereby also gathering and spreading knowledge throughout the region. Thus it was the cultures mingled, ushering in a period of growth and development.."
(http://gorp.away.com/gorp/location/africa/phonicia.htm)

Later in the next Millennium, the Roman Empire expanded their political and commercial boundaries by developing roads and bridges (the second layer in the platform for commerce) as well as developing aqueducts to deliver water to promote regional growth and trade. Subsequent layers of railroads, telephone networks, power grids, and airports followed over a Millennium later.

Trade Routes Have Become Electronic

Expanding trade routes and establishing commerce means the application of creativity, technology and motivation to overcome natural obstacles including water, land and air which over time were all conquered. Today, those trade routes have become electronic and photonic.

Throughout the last five Millennia, developing new trade routes has been important to the expansion of trade, culture and commerce. Now, those trade routes have evolved into electronic highways in the latest layer of critical infrastructure implementation.

Broadband connectivity is the latest layer of critical infrastructure. Having this single layer providing an intelligent amenity to a commercial building or campus can eliminate 90% of the competition in most real estate markets. New real estate strategies must be forged and implemented. This will not remain that large of a competitive advantage for long, so the time for action is now.

THE TRANSPORTATION OF INFORMATION

The importance of the high-speed internet is finally being recognized in this millennium by those who should have been rebuilding their copper "roadbed" of telephone network they built in the United States decades ago.

Just like single-lane dirt roads which evolved into the multi-lane superhighways of today, the single-function copper-based voice network has to be updated and replaced to a multi-channel, multi-gigabit, fiber-optic based network which can handle the explosive growth of wireless, video and other convergent applications. This also includes updating wireless network infrastructure broadcasting those signals to handheld devices.

Just as you can't drive fast on a dirt road, you cannot speed fast over copper. As I have written in various articles discussing the shortcomings of a tired network infrastructure, *"putting DSL on copper is like putting on a vinyl top on a stagecoach in the era of the space shuttle."*

Trying to compete in a global economy requires a much faster network with dynamic allocations for supporting new video-based applications in both terrestrial and wireless transmission media. As the rapid proliferation of Smartphones and Tablets continues, you will see more bandwidth-hungry applications drive the market and raise the bar for baseline speeds for network infrastructures in every country.

Multi-gigabit speeds will be replaced by terabit speeds in the near future. With traffic growing exponentially, terabit speeds will become a reality within the backbones of public networks within a decade.

With the tens of millions of dollars that incumbent phone companies in the United States spent in early years after 2001 to lobby for competitive restrictions which amount to a phone network "protectionism", they could have been much further ahead into their endeavor of implementing a next-generation network that could handle much higher rates of speed which will be required by 2020.

ARE WE AT THE END OF THE EMPIRE?

Contrary to what many say about the United States being in the forefront of network technology, the reality is that we are more in the catch-up mode when it comes to implementing broadband connectivity. Some countries like South Korea have surpassed us when it comes to network infrastructure and broadband connectivity.

There is a significant price-tag on renewing all layers of our infrastructure. It is not enough to implement a new roads program or a bridge and dock program. Any comprehensive infrastructure study which looks at what is needed in any state must include reviewing the state's layer of network infrastructure for broadband connectivity.

If they don't address and fund this layer of infrastructure, they might as well forget about attracting and maintaining leading-edge organizations. These organizations already have broadband connectivity as a major "site criteria" when they look for areas to establish new corporate facilities.

As they say, if we do not heed the mistakes made in the past, we are condemned to repeat them in the future. We cannot afford to let any portion of the United States lag behind in this redefinition of the framework of infrastructure.

Cities and regions' leaders as well as economic development teams need to understand their infrastructure can either attract or repulse new corporate facilities. They also need to understand when they talk about investing in infrastructure, by adding or improving upon it, they need to address the whole framework of infrastructure. They need to be more encompassing and include all layers of infrastructure and not just the "traditional view" which includes roads and bridges, public transportation (rail) and highways.

I call this the "1950s infrastructure vision" which was focused on building more infrastructure to support the first three levels which include driving on roads, developing more highways, and from a metropolitan standpoint, adding more stops on the rail lines.

CARLINI-ISM: *"1950s solutions will not solve 21st century economic problems."*

Economic development means creating jobs and jobs keep regions viable. When jobs leave a region, increased crime and drugs move in to replace them. When people are working, tax revenues keep municipal services alive and vibrant. When tax revenues go down, crime and more social programs go up.

CARLINI-ISM: *"Good government fosters good commerce. Bad government fosters no commerce and attracts crime."*

All municipalities need to review what they have as far as in place infrastructure and analyze what needs to be done in order to build a solid platform for commerce to build upon. They need to re-vitalize their infrastructure, including network infrastructure and not think like it is the 1950s. This is what many business writers, economic commentators, and traditional thought leaders on various industry perspectives are missing: old solutions do not solve today's problems.

Traditional city politicians, economic development consultants, and the traditional business and municipal advisors of lawyers and accountants also do not get this concept. Most are still thinking and acting within a 1950s mindset.

A revitalization of the first three layers of infrastructure defined in the global ***Platform for Commerce*** (**See CHART 5-1**) does not address the broadband connectivity needs of today and tomorrow.

In today's post-information age, if organizations fail to incorporate their communications based information systems strategy as a platform for their overall business strategy, they might as well have no strategy. And, they won't achieve "Best Practices" either.

Achieving "Total Quality Management" or in today's buzz phrase: "Best Practices" means knowing how to apply technology. And, "Best Practices" are a moving target to aim at. What worked last year could be completely obsolete this year. This includes network topographies and baseline speeds within those networks.

CARLINI-ISM: *"Best practices are a moving target. Best practices need to take into account applying technology. And, best practices change with the weather."*

WHAT IS THE RECIPE FOR SUCCESS?

In Silicon Valley, you have people working together for a common cause. Venture capitalists, entrepreneurs, bankers, politicians, academia and most importantly, media, make up the "recipe for economic success." **(See CHART 5-2)**

CHART 5-2: THE RECIPE FOR ECONOMIC SUCCESS

(All of these entities have to mix and move in the same direction.)

Entrepreneurs	The "catalyst" of the IDEA
Angel Investors, Venture Capitalists	The beginning financing
Engineers, Salespeople	The builders, customizers and sellers of the products
Bankers	The sustaining financing
Politicians	The source of grants and other help
Academia	The source for ideas and future workforce
Real Estate Developers	The builders of the platform for the business
Media	The glue that ties everyone together (the secret ingredient)

Media? Yes, that is the secret ingredient for regional success.

In Silicon Valley, they have people in the media who understand the technology as well as the total picture. And as journalists and reporters, they can articulate and spotlight new trends and developments which create a larger and larger following nationwide, if not internationally.

Through their reporting, they attract a broader, interested source for venture capital funding as well as a demand for action.

CARLINI-ISM: *"He who controls the press, controls the rest."*

In every regional area, if you want to understand if they are maximizing their potential infrastructure, just tune into their local television and radio stations to hear how their news teams and reporters discuss technology and its importance to the region. If all you hear are fluff pieces and stories which are not in-depth enough to reflect the impact of technology onto the businesses and economy, chances are, there is not anything of significance going on. And if there is, the reporters do not know how to report it because they simply don't understand it.

My white paper, *"INTELLIGENT BUSINESS CAMPUSES: KEYS TO FUTURE ECONOMIC DEVELOPMENT"*, published in 2008, by the International Engineering Consortium), discussed the new approach to Master Planning for new developments. One of the new key concepts is the idea of including planning for network and power requirements upfront, before the road is built, the parking lot is laid and the landscaping is planted. This not only saves a lot of money, but it also helps to more clearly define the *Platform for Commerce*.

Many traditional real estate and business park developers have yet to adopt this approach and continue to build for the wrong century.

Thinking about where utilities should be brought in and how much capacity should be available actually saves money on a project. This extra planning step adds significant value as the target market for tenants becomes more sophisticated and those regions that can offer more will attract a higher level of potential tenants.

In today's markets, you want to stay away from the lowest common denominator "space available" market strategy, unless you want to sell a commodity.

ANTIQUATED UTILITIES APPROACH

Horse-and-buggy approaches to commercial real estate which include: one connection to one central office of the phone company for network services and one connection to one power company from one station on the power grid are obsolete. Get rid of these obsolete rules-of-thumb immediately because they do not support mission critical applications of today's corporate core businesses. **(See CHART 5-3)**

Double the connections as well as the carriers on each of these two intelligent amenities (power and broadband connectivity) and you just have ascended into providing a more rigorous, resilient platform for a whole new tier of sophisticated corporate facilities that site selection committees are looking for. This is well above-and-beyond today's status quo in real estate. By doing this, you have effectively cut out at least 95% of your competition when it comes to seeking out a Class A tenant (more likely 98%). And, this improvement is not going to break your budget for these "access to infrastructure" improvements. **(See CHART 5-4)**

CARLINI-ISM: *"Best practices are not found in bureaucracies."*

CHART 5-3: UPGRADING INTELLIGENT AMENITIES

INTELLIGENT AMENITY	OLD WAY	NEW WAY
Network Communications	Single line, single route, single carrier	Multiple line, multiple route, multiple carriers
Power	Single line, single route, single carrier	Multiple line, multiple route, multiple carriers
Management of HVAC Systems	Locally managed, within building	Remotely managed, can gain some economies of scale if several buildings are run on one centralized system.
CONTROLS	Single zone, maybe floor-by-floor zones	Multiple zones within one floor. More individual and dimming controls.

Address the whole *Platform for Commerce* when thinking about improving the region. If a new business campus or industrial park is being suggested as part of a regional renewal program, make sure it can get redundant power and broadband connectivity to it. You just might be creating a real opportunity for some corporate facilities to be built on it. New facilities equate to new jobs and new jobs equate to a stronger tax base.

CHART 5-4: INTELLIGENT AMENITIES

POWER	CONNECTIVITY	DESIRABILITY	PROPERTY AVAILABILITY
SINGLE SOURCE	SINGLE CARRIER	VERY LOW	VERY HIGH
SINGLE SOURCE	SINGLE CARRIER (multiple routing)	LOW	HIGH
SINGLE SOURCE (two separate grids)	MULTIPLE CARRIER (diverse routing)	BETTER	LOW
DUAL SOURCE	MULTIPLE CARRIER (diverse routing)	HIGH	VERY LOW
MULTIPLE SOURCE (including Alternative Energy)	MULTIPLE CARRIER (including both wired and wireless carriers)	VERY HIGH	EXTREMELY LOW

CHAPTER REVIEW

THINK ABOUT THESE CONCEPTS AND QUESTIONS BEFORE MOVING INTO THE NEXT CHAPTERS.

Understand the *Platform for Commerce* and its importance to regional sustainability.

What is the **Recipe for Success** for regional economic development?

What has changed in **Master Planning** for corporate campuses and industrial parks?

CARLINI-ISMS

"'Building for the future'" is a very hollow statement, when funding is aimed towards maintaining the past."

"To better see into a complex future, a clear framework is needed to structure new concepts."

"1950s solutions will not solve 21st century economic problems."

"Good government fosters good commerce. Bad government fosters no commerce and attracts crime."

"Best practices are a moving target. Best practices need to take into account applying technology. And, best practices change with the weather."

"He who controls the press, controls the rest."

"Best practices are not found in bureaucracies."

[6]

INTELLIGENT INFRASTRUCTURE: THE PAST

"When I want an opinion, I'll get it from my peers - from men of vision, like our great railroad builders... Stanford, Huntington, Dinsmore... Fellows with imaginations broad enough to span the continent." – **JOHNATHAN RABEN**

Making decisions on infrastructure can affect regional viability for years to come. In the past, certain key decisions on something as simple as letting the railroad come through the town meant a huge economic boost to a region. You would think getting that added boost, getting that "extra layer of infrastructure" would be regarded as a smart decision. In some cases, the right decision was not made and you can see its negative, residual results across future generations.

Let's go back to when the railroad was first being introduced as the next layer of infrastructure in the United States which would expand trade routes. Not everyone was keen on the idea, including those who saw the railroad as an economic threat as well as a big competitor to their business after the Civil War.

This is a good story about increasing regional viability and economic growth. Today, the strategy of increasing trade and commerce hasn't changed, only the tactics and what should be implemented has changed.

Political Decisions Can Affect Future Regional Viability

Let's go back just after the Civil War in 1865 when the framework for the **Platform for Commerce** looked like this and the "newest" layer of critical infrastructure was the railroads **(SEE CHART 6-1)**:

CHART 6-1: LAYERS OF THE PLATFORM FOR COMMERCE (circa 1865)

3	RAILROADS	UNITED STATES (mid-1800s)
2	ROADS/BRIDGES	ROMAN EMPIRE (500BC- 476AD)
1	PORTS/ DOCKS/ WATER	PHOENICIANS (1200BC-900BC) EGYPTIANS (3000BC-1400BC)

Source: JAMES CARLINI

The last big infrastructure issue facing Chicago and St. Louis back in the late 1800s after the Civil War was the approval of the railroads to get rights-of-way for a centrally-located hub. As they built the Transcontinental Railroad from the East Coast and the West Coast, the railroads wanted to select a central hub that different tracks could meet.

At that time, St. Louis was considered a more important city than Chicago and was actually a larger city by about 30,000 in population. It was the well-established "jumping off" point to the Western territories of the United States for wagon trains from the East headed to California and other points west. The city's motto is still *"Gateway to the West.'* The railroads of the day wanted to capitalize on St. Louis' central location and establish their regional hub there.

From the influence of the riverboat industry, which controlled transportation of goods on the Mississippi and Missouri Rivers, St. Louis restricted the railroads' access into St. Louis and crossing the Mississippi River. It was "the right thing to do" – or so they thought.

On the other hand, when the railroads approached Chicago politicians to get permission to build the central regional hub in Chicago, they let the railroads have access to terminating into Chicago as a regional transportation hub. With that decision, Chicago grew faster in the next 25 years because they had an additional layer of infrastructure which catapulted their influence on regional economic development as well as their population.

The chart below shows the unmistakable, long-term impact that restricting infrastructure development has on the regional economy (**See CHART 6-2**). St. Louis never recovered from their short-sighted decision to restrict the railroad.

Its politicians failed to understand the importance of adding the next layer of infrastructure into their region for economic growth and regional sustainability. The results? Economic stagnation and a long-term loss in growth and expansion.

CHART 6-2: POLITICAL MISTAKE ON INFRASTRUCTURE DECISION

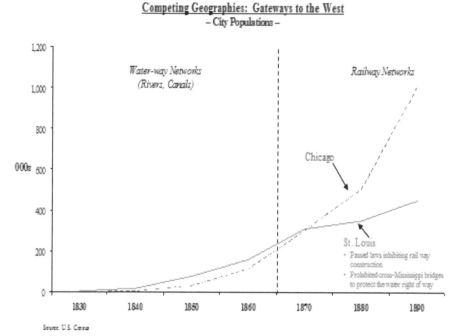

Competing Geographies: Gateways to the West
– City Populations –

Did you know Abraham Lincoln made a lot of money in some of the negotiations with railroad rights-of-way? Lincoln was one of the lawyers who helped construct some of the original agreements for the railroads in the region. In Decatur, Illinois where he first practiced law before becoming President, there is a small county museum with many of Lincoln's early legal documents and other memorabilia.

The curator of the museum said that even though Lincoln filed a lot of records in the county, most of them are no longer there because of people taking the documents with his signature on them.

It is an interesting observation to note that on a recent drive along the Mississippi River in Illinois, railroad tracks parallel the river for miles. It is like the railroads made sure to get back at the riverboat captains who pushed the politicians in St. Louis to restrict railroads from coming into the city. Whatever route the Mississippi River provided for trade and moving goods was duplicated by the railroads running tracks in parallel to the river itself. The railroads did provide competition against the riverboats and the beneficiaries were all the consumers.

What if the railroads and riverboat captains worked together instead of fighting each other? If the riverboat captains thought about making some type of major dock and drop-off point at the port of St. Louis, they could have created a major distribution point where the railroads could have taken their cargo. The railroads could have extended the riverboats distribution to outside the waterways and out to the West. Evidently, no one was looking at working together at the time.

If this "collaboration" happened between the two different layers of infrastructure, St. Louis would have become the major Midwest regional hub instead of Chicago, Illinois. This would have changed the whole economic landscape going into the early 1900s.

This is one thing which seems to get lost in many arguments about restricting infrastructure and creating regulatory obstacles for adding or modifying infrastructure. Competition is good. Monopolies are bad. This applies to all of network infrastructure as well.

CARLINI-ISM: *"Competition is good. Monopolies are bad. Having competitors accelerates innovation within an industry."*

When you don't have enough competition, prices can be set and there is no reason to innovate and cut costs or make things more efficient because there is no one competing for the same market. When you have competitors, it seems to accelerate innovation and keep prices relatively fair.

This is not a book on economics or regulatory affairs, but it should be noted that the network infrastructure evolved from a very big monopoly in the United States, the Bell System, which was then divested in 1984. Since then, the network infrastructure has supposedly been a more competitive environment with competitors trying to entice various segments of the market.

In some respects, people would argue that the incumbent phone companies are still trying to reap the benefits of the restrictive practices of a monopoly, even in a market that is supposedly open to competitors.

The incumbent phone companies are not favorable to competitors coming in with more state-of-the-art infrastructure. We need to get that bottleneck changed. It seems like we have another feud between the incumbents and any other entity wanting to build out more infrastructure. Is history repeating itself with the incumbent phone companies trying to restrict outside companies from building up new infrastructure that can add to the ***Platform for Commerce?***

Don't Let History Repeat Itself

As they say, if we do not heed the mistakes made in the past, we are condemned to repeat them in the future. We cannot afford to let any portion of the United States lag behind in this new 21st century global economy. This is important when we require the latest layer of infrastructure, broadband connectivity, to be accessible to all and at a reasonable price. It is much better to have this layer within the region, than to restrict it or not employ it.

Remember the City of St. Louis' ill-fated decision on the railroad. Even more importantly, remember their lack of keeping up with regional growth and falling behind in economic development because they were lacking that extra layer.

A commitment of revitalization of the first three layers of infrastructure defined in the *Platform for Commerce* (See **CHART 5-1**) does not address the broadband connectivity needs of today and tomorrow.

CARLINI-ISM: *"1950s solutions will not solve 21st century economic problems."*

All municipalities need to review what they offer as far as infrastructure and analyze what needs to be done in order to build a solid platform for commerce to build upon. They need to re-vitalize their infrastructure, including network infrastructure and not think like it is still the 1950s. This is what many business writers, economic commentators, and traditional industry perspectives are missing.

Traditional city politicians, economic development consultants, and the traditional business and municipal advisors of lawyers and accountants also do not get this concept. Most are still thinking within a 1950's mindset. Those executives running many of the utility companies are also lacking the motivation to look at constructive change and creating new policies even though technology is driving change within each of their industries.

CARLINI-ISM: *"Best Practices are a moving target. And, best practices change with the weather."*

CRITICAL TRADE ROUTES ARE NOW ELECTRONIC

Since the railroads crisscrossed America and tied the United States together, other layers of infrastructure were developed, built, and added onto the **Platform for Commerce.**

New modes of transportation defined and expanded new layers within our **Platform for Commerce** and they provided residual benefits to the regions where they were employed.

The airplane was a significant accomplishment in the very beginning of the 20th century. Within a little more than a century of time, it went from an experimental vehicle and initial military vehicle to a very sophisticated form of commercial and freight transportation by air. Use of the airplane created all new fast, non-stop trade routes where none existed before or existed within a lower level of the **Platform for Commerce.**

Within a couple of decades of its inception, the airplane became a new way to cross continents and get to destinations which may have been totally inaccessible. Steamship lines which were the elite way of passage over oceans for the upper class saw their importance dwindle as airplanes got more powerful, more sophisticated, and became the new cost-effective transportation mode of choice when going long distances.

New trade routes were established by air. These routes became accepted and grew in their importance as well as in their capacities to transport people as well as commercial cargo. Airports became a key economic driver for cities to implement, if the cities wanted to compete with other cities for tourism, commercialism, and other economic development areas. New products were derived from the ability to fly in something fresh which could have never happened within the prior layers of the **Platform for Commerce**. This new "overnight delivery" ability created brand-new businesses and micro-industries, from tropical flowers grown in Hawaii to fresh fish flown in from either coast.

Today, trade routes have become electronic and photonic. With the Internet providing a whole new layer of communications and "intelligent infrastructure", more trade routes were established and more destinations developed on an electronic infrastructure as well as the traditional "bricks-and-mortar" infrastructure. Again, this will set the groundwork to establish and maintain new businesses and industries as well as new products which would not be able to flourish within the previous layers of the **Platform for Commerce.**

CARLINI-ISM: *"Network infrastructures are like icebergs. You only see 5% of the issues and responsibilities on the surface."*

OTHER LOCAL AND GLOBAL INITIATIVES

Canada is developing a huge multi-function logistics center in Winnipeg. A huge airport and logistics center covering over 20,000 acres, named Centreport, is currently being developed. Their goal is to try to create a whole new hub of regional economic development and a whole new destination for international commerce deep within their country. (In effect, an inland port open for international trade.)

Canada sees a great opportunity with their "Inland Port" concept. They will create and sustain a lot of jobs in an area that does not have a lot going for it right now. They are building a 21st century ***Platform for Commerce.***

Other municipalities and regional locations are focusing on upgrading different layers of their infrastructure because they see the importance of offering a solid ***Platform for Commerce*** to those organizations which can bring new jobs, new capabilities, and broader economic development to a region.

Intelligent buildings are now clustered into Intelligent Business Campuses which require a new layer of infrastructure to compete in a global economy. The new layer is broadband connectivity. When it comes to broadband connectivity as a critical layer of infrastructure, Fort Wayne, Indiana, is a good example to review. We can see some of the residual impact of adding fiber optics to connect the entire city has resulted in a stronger economy with businesses starting to come back and establish corporate facilities there.

Speaking to Fort Wayne's former mayor, Graham Richard, several years ago, he said Fort Wayne entered a contract with Verizon to add a huge amount of fiber optic cables into the Fort Wayne topography. At the time, the price tag for the infrastructure upgrade was over $110 Million. Fort Wayne (about 250,000 in population) became the first city in the Midwest to utilize FIOS from Verizon.

There has been a positive return on their investment in upgrading their network infrastructure. Their shift to fiber-to-the-premise (FTTP) in order to attract new growth in jobs and businesses shows it has paid off.

Fort Wayne's Richard, who was touted as "America's Broadband Mayor" for his efforts to get Fort Wayne some real fiber optic connectivity, pushed an agenda to get the job done and was rewarded by his efforts through more companies committing to opening up facilities in Fort Wayne. Big bandwidth coming in on FTTP (Fiber to the Premise) became available in Fort Wayne for anyone from a corporate facility to an individual working from their house.

My question when designing corporate networks has always been: *What if bandwidth was not an issue?* What applications would we then launch? In so many past endeavors for new corporate applications, the addition of the application to the enterprise network is voted down because "there is not enough bandwidth on the organization's network." What if we could get rid of that planning bottleneck?

Getting more bandwidth opens up new directions and new applications which before were viewed as impossible.

CARLINI-ISM: *"Bandwidth is like garage space. The more bandwidth you have, the more you will fill up."*

Bandwidth is like garage space. Owners of a 2-car garage wish they had a 3-car garage. When they get a 3-car garage house, they begin to think, *"If I only had a 4-car garage."*

Whatever space you get, you will find a way to use and fill it up. The same goes for bandwidth. The more bandwidth we provide to both corporate entities and individual users, the more creativity they will use in order to utilize the bandwidth for new endeavors.

In the past, outside plant engineers for the Bell System designed underground cable capacities putting enough in the ground so they would not have to go back and add to it for a good 20-25 years. Traffic back then was fairly stable and predictable as most of it was voice traffic. As the years went by, we have added a lot of data as well as video traffic to that growth and now we are adding applications to handheld devices (Smartphones and Tablets). Growth in traffic has caused us to re-evaluate design concepts and consider new growth rates for designing network capacities.

About thirty years ago, a 9600 bit per second network (9.6Kbps) was considered a "fast" network speed for corporations.

One modem for a corporate network cost about $7,000 (circa 1982). One point-to-point circuit requires two modems (one at each end). Imagine having to pay $14,000 just for the modems to run one 9.6Kbps data circuit.

How much do digital service units cost today? A couple of hundred dollars to run multi-megabit speeds (if that)? Speed-wise we have come a long way since the Divestiture of the Bell System (1984).

Looking forward, those network speeds will accelerate even faster. We need to insure we are building networks to take us into the future and not build something which needs to be replaced every couple of years. Some network planners are way off in their rules-of-thumb.

Some people talk about putting in T1 circuits (1.544Mbps) to their enterprise as if digital T1 circuits are the "new technology." Digital T1 Circuits have been around since the early 1960s.

CARLINI-ISM: *"The first T1 circuit was installed in Skokie, Illinois in 1963. Hardly, cutting-edge technology today."*

If anything, many network engineers today are under-engineering networks and then they are going back to re-engineer them the very next year. (There are enough examples in the NFL and other Sports stadiums where they missed the ball as far as capacity.)

We need to insure we are building good networks which can support long-term and explosive growth as more people turn to wireless devices as a ubiquitous edge technology. Too many are still adhering to obsolete rules-of-thumb. **(See CHART 6-3)**

Understanding broadband connectivity and the impact of high-speed transmission as an intelligent amenity is critical for any elected politician. Municipalities have to understand the importance of electronic trade routes and network infrastructure as well as understanding the more traditional layers of infrastructure like roads, bridges, and railroads.

When Class A Buildings slip to Class B offerings because they are technologically obsolete within today's markets, the segments of the market which they can serve and compete for, shrinks. Technical obsolescence should negatively affect the appraisal of a building's value and the elements defining the facets of technology should be understood by those in the real estate markets.

CARLINI-ISM: *"Don't say you're "state-of-the-art" if you are still using horse-and-buggy rules-of-thumb for your building(s)."*

CHART 6-3: OBSOLETE BUILDING RULES-OF-THUMB

BUILDING ACCESS	TRADITIONAL PHONE & POWER COMPANIES RULE-OF-THUMB
POP (Point-of-Presence)	One connection into the building
CONNECTION TO TELEPHONE COMPANY's CENTRAL OFFICE	One connection and one route (no diversity)
AMOUNT OF SERVICES (CABLING)	1 Pair for every 200 square feet, 1 Pair for every 300 square feet
CONNECTION TO POWER COMPANY's SUBSTATION	One connection and one route (no diversity)
POWER	5 Watts per square foot
CORPORATE NETWORK SPEEDS (Circa 1982-1983)	9600bps (9600 bits per second)

CHAPTER REVIEW

CARLINI-ISMS

"Competition is good. Monopolies are bad. Having competitors accelerates innovation within an industry."

"1950s solutions will not solve 21st century economic problems."

"Best Practices are a moving target. And, best practices change with the weather."

"Network infrastructures are like icebergs. You only see 5% of the issues and responsibilities on the surface."

"Bandwidth is like garage space. The more bandwidth you have, the more you will fill up."

"The first T1 circuit was installed in Skokie, Illinois in 1963. Hardly, cutting edge technology today."

"Don't say you're 'state-of-the-art' if you are still using horse-and-buggy rules-of-thumb for your building(s)."

[7]

INTELLIGENT INFRASTRUCTURE: THE PRESENT

"Economic development equals broadband connectivity and broadband connectivity equals jobs." – JAMES CARLINI

A revitalization of the first three layers of infrastructure defined in the *Platform for Commerce* **(See CHART 7-1)** does not address the broadband connectivity needs of today and tomorrow.

When the stimulus packages came out to try to revitalize the US economy back in 2010, it focused more on the first three layers of the *Platform for Commerce*. Not much money was ear-marked for other infrastructure levels.

A better expenditure of funding for the 2010 and 2011 stimulus packages would have been to get all those who had computer skills and hire them to work in larger project software jobs, like the Affordable Care Act (ACA) project as well as others.

The present approach to planning and implementing intelligent infrastructure is not being practiced by everyone. Many are still stuck in a 1950s mentality. There needs to be a whole new generation of education in real estate when it comes to understanding and applying concepts in a multi-disciplinary, strategic approach for next-generation business campuses as well as stand-alone intelligent buildings.

CARLINI-ISM: *"20th century solutions will not solve 21st century problems."*

Remember in 2010, people in the United States talked about funding "Shovel-ready projects?" Unfortunately, "shovel" skills were not the skills needed to re-energize the economy. If a project was focused on "shovel ready", it was focused more on providing work for those with 1930-1950s skill sets (again the Rote, Repetition and Routine crowd) than those who were actually out-of-work. (IT professionals, engineers, and other skilled and degreed people)

What was needed were more "Keyboard ready" projects which would have engaged more people who were actually out-of-work and who were sitting out on the sidelines with underutilized, cutting-edge skills. These people were either idle or underemployed with software development and engineering skills, "Keyboard ready" projects should have been the projects created.

The Affordable Care Act (ACA) which came about in the same time period would have been a good, challenging project. In order for the ACA to be a viable program, large complex systems were needed to be redefined and overhauled in both federal and state programs.

CHART 7-1: THE PLATFORM FOR COMMERCE

LAYER	LEVEL	DOMINANT INITIAL DRIVER OF IMPLEMENTATION IMPORTANCE
SPACE (INTERPLANE-TARY) (FUTURE)	8	JUST BEGINNING TO BE BUILT (Space shuttles, space station, satellite networks) (Future: mid-21st century, 22nd century? US, RUSSIA, JAPAN, CHINA?)
BROADBAND CONNECTIVITY NETWORK (CYBER-INFRA-STRUCTURE)	7B (wireless) 7A (wired)	CHINA, JAPAN, S. KOREA, NETHERLANDS, US (beginning 21st Century, IBCs, IIPs & IRECs)
AIRPORTS	6	EUROPE, UNITED STATES (mid-20th Century)
POWER (GRIDS, NUCLEAR POWER, OIL)	5B (Nuclear) 5A (everything else)	UNITED STATES (beginning/ mid-20th Century)
TELEPHONE NET-WORK (ANALOG VOICE ONLY)	4	UNITED STATES (beginning/mid-20th Century)
RAILROADS	3	UNITED STATES (mid-1800s)
ROADS/BRIDGES	2	ROMAN EMPIRE (500BC- 476AD)
PORTS/ DOCKS/ WATER	1	PHOENICIANS (1200BC-900BC) EGYPTIANS (3000BC-1400BC)

Source: JAMES CARLINI, 2009, 2013. All Rights Reserved

Re-doing this whole area would have employed many out-of-a-job who had those necessary "systems engineering" and software development skill sets. Discussing the merits of this government project creating a lot of jobs should be dealt with within another book.

There are just too many facets to write about and the discussion would spill over into another book. One chapter would not do the subject justice. It does demonstrate we could have re-employed many people into skilled jobs and would have strengthened many local economies while adding to the higher layers of the **Platform for Commerce**. With more large systems work being developed, the network infrastructure would also have to be improved and it could have been a good residual investment from the Stimulus package. We need to identify and take advantage of those types of large opportunities, especially if they are embedded within the layers of the **Platform for Commerce**. One thing that would become clear, any project focused on expanding the infrastructure would have a good payback immediately, as well as, into the future.

We need to assess infrastructure as an all-inclusive look at ALL levels of the **Platform for Commerce** and not just the three or four layers some people incompletely visualize when discussing infrastructure. We need to see the total picture and understand the total economic impact of the framework of the **Platform for Commerce** when looking at projects as well as how we assess all the issues concerning a new endeavor.

Let's go back to the post-Civil War era for a second. What if the competing factions of railroads and riverboat captains collaborated, instead of fought each other? What if the riverboat captains said, *"Let's set up a distribution point in St. Louis where we can unload our goods and the railroads can take them out West?"*

How different would the transportation infrastructure develop if they had worked together instead of trying to restrict the two layers of infrastructure at that time?

How much faster would it have been to develop an integrated, two-level transportation system using both the riverboats and trains to deliver to new territories and towns? What if they initially worked together instead of working against each other?

These types of questions can be raised again when we discuss the importance of today's incumbent phone companies being more open to competitors rather than fighting them for market dominance. We must also become more cognizant of converging disciplines and the new collaboration needed between the areas of real estate, technology, and infrastructure.

How Do You Assess Real Estate Today?

Today, you must look at real estate projects as a more complex endeavor. It is not enough to understand real estate, you need to recognize it is converging with other large elements: the infrastructure, technology (both communications and information systems architecture) and regional economic development. Incorporating and energizing every element is important in insuring success of the project and providing a facility that can support 21st century businesses.

Each element's specific attributes will determine the shape and outcome of the project:

REAL ESTATE: What and where is the land that the project will be on? Is it the best use of the land? Are there any issues/obstacles to be aware of? What government incentives are connected with the land parcel? ((Be sure to check to see if there are any TIF (Tax Increment Financing) districts or other incentives tied to the property))

INFRASTRUCTURE: What is in-place/ available/ accessible to that piece of land? Access to transportation? Rail? Airports? Access to power and network communications access (broadband connectivity – both wired and wireless)? Also find out if anything is planned for the near and long-term future.

TECHNOLOGY: What extra intelligent amenities have been added onto the infrastructure to make it more desirable? Redundant network carriers providing Multi-Gigabit connectivity? Redundant power suppliers? Wireless networks to handle high capacities of Smartphones? Other cyber/connectivity intelligent amenities?

ECONOMIC DEVELOPMENT: How is this project being positioned within the region? How is it being packaged and promoted to prospective corporate tenants? How is it being funded in both its initial and operating phases? Who is selling its viability, not only for the region but for the people themselves? **(See CHART 7-2)**

New projects are one thing, retrofitting existing buildings and campuses represent a whole other facet of skills in building and re-building a region's ***Platform for Commerce***. You have to determine if a building is technologically obsolete. Several requirements have changed over the years and many buildings need to upgrade in order to be able to compete for today's technology-dependent tenants.

At this point with mission critical applications and their supporting networks being utilized in every enterprise, you should not be in a building which only offers a single connection to a single network carrier. Not having redundancy of connectivity to the central office creates a single point-of-failure and negates any claim of having "mission critical" capabilities for a potential corporate tenant.

CHART 7-2:   The RITE Approach to Real Estate Projects

THE PACKAGE FOR REAL ESTATE PROJECTS

ELEMENTS	ATTRIBUTES
R EAL ESTATE	LAND
I NFRASTRUCTURE	IMPROVEMENTS (WITHIN & SURROUNDING THE LAND)
T ECHNOLOGY	INTELLIGENT AMENITIES
E CONOMIC DEVELOPMENT	MARKETING/ PROMOTION/ FINANCING/ PACKAGING

Source: ©James Carlini, Copyright 2012

BUILDING CONNECTIVITY: FIVE WAYS TO TELL IF YOUR BUILDING IS OBSOLETE

Some will argue about the dual connection to two separate telephone company central offices being necessary, but when you look at the growth in mission critical applications, you need to have this diverse connectivity to support the redundancy of the network.

Today, one out of every three applications are considered mission critical with the amount growing to one out of every two applications within several years – or less. My questions to the "experts" are:

- If you engage mission critical applications in your organizations, how can you still be connected to the central office via one route?

- How can you run the electronics from only one power source? (Forget battery back-up for a couple of hours, the design concept should be "business continuity" not "disaster recovery") You need to implement separate and diverse power sources along with separate and diverse routes into the network infrastructure if you are going to support mission critical applications. Anything less, is unacceptable.

- What about your applications running on your enterprise network?

- Are you getting the speeds you need to compete in the 21st century? Or, are you stuck with 20th century connectivity because no one knows how to upgrade the building's network infrastructure? Or, does no one want to make an investment to upgrade the existing infrastructure because it is an investment?

- How receptive is your building to all the new connectivity technology exploding onto the market? This is a question becoming more common to those trying to make sure they lease the right space for their organizations.

- Do you have your Smartphone with you? Does it work well in your building? Can you download streaming video and get a good video, or is the picture choppy? If not, most will opt out and check out another property. This is happening today.

CARLINI-ISM: *"A Smartphone application isn't any good if you cannot access it."*

More questions need to be asked as organizations keep adding new applications to run their businesses:

- What about your applications running on your enterprise network?

- Are you getting the speeds you need to compete in the 21st century?
- Or, are you stuck with 20th century connectivity because no one knows how to upgrade the building's network infrastructure?

These are some of the questions you must answer when you begin to assess the intelligent amenities available within an existing building. Comparing buildings is becoming more than just which one has the nicer lobby or the faster elevators.

Buildings which do not support access to broadband connectivity or redundant power sources are becoming technologically obsolete.

CHECKLIST

If you don't have separate connections to the central office, your network is your single point-of-failure within your enterprise application. The same applies to power. Having two separate sources for these intelligent amenities makes a building more attuned to what is needed today and tomorrow. Otherwise, how can it be considered to be able to support "mission critical" applications?

Here is a quick checklist to assess a building's network infrastructure: **(See CHART 7-3)**

CHART 7-3 NETWORK INFRASTRUCTURE CHECKLIST

WHAT TO CHECK	WHAT YOU MAY FIND	WHAT IT SHOULD BE
CONNECTIVITY TO THE CENTRAL OFFICE (CONNECTION)	A SINGLE CONNECTION TO A SINGLE CENTRAL OFFICE	TWO SEPARATE CONNECTIONS TO TWO SEPARATE CENTRAL OFFICES
CONNECTIVITY TO THE CENTRAL OFFICE (Type of transmission media used)	COPPER	FIBER OPTIC
WIRELESS CAPABILITY (Any WiFi or DAS?)	Standard network carrier coverage.	Multiple network carrier coverage. (PLUS – capacity, not just coverage)
*TWINS * (See below)*	A single central office connection with a small amount of spare capacity in it as well as the vertical riser system (or maybe no spares).	It should have spare capacity both to the central office as well as the vertical riser system and be able to handle gigabit speeds.
FIRESTOPPING	Many penetrations where cable is pulled through the floor or wall that is NOT fire-stopped (covered with flame retardant material to stop smoke from spreading across the building.	ALL penetrations between walls and floors should be fire-stopped with materials made for that purpose. It is more a life/safety issue than a connectivity issue, but you should be aware of it.

A single connection to the building from a single central office was the approach for over a century. It was literally since we had stage-coaches, but we are beyond that if we are concerned with supporting mission critical applications. Horse-and buggy rules-of-thumb for communications cannot be applied in 21st century applications.

What Are Twins

Years ago, I wrote a rule-of-thumb for the cabling communications industries, which is still used as a way to understand what network infrastructure is coming in and being used in a building. It was also used as a teaching tool to understand what cabling capacity was coming into a commercial building. With the TWINS © tool, you can easily figure out what the copper capacity is within a building. **(See CHART 7-4)**

TWINS © stands for
TOTAL PAIR,
WORKING PAIR,
IN-SERVICE PAIR,
NON_WORKING PAIR, and
SPARE PAIR

For example, in a building that has a total of 4,000 cable pair coming into it from the central office, how do you determine what amount of them are actually being used and how much is actual spare capacity? **(See CHART 7-4)**

CHART 7-4: TWINS – BUILDING CABLING EVALUATION

CODE	Type of PAIR	TOTALS
T	TOTAL PAIR	4,000
W	WORKING PAIR	3,300
I	IN-SERVICE PAIR	2,200
N	NON-WORKING PAIR	700
S	SPARE PAIR	1,100

How many building owners and property managers even know what cabling capacity they have in their buildings? If there is no record of what is available and what is in-service, how can they lease out a building if they do not know how much spare capacity of connectivity the building has? Would you move into a building where you could not expand or add on more communications capacities as your organization needs them? How can you manage something if you don't understand how to measure it?

These questions were not as critical in the past, but are very critical today. Insufficient cabling to and within a building will degrade its marketability and lower its value. Anyone paying top dollar to buy or lease an obsolete building is a fool because it takes millions of dollar to bring it back up to speed. (No pun intended)

CARLINI-ISM: *"In today's commercial building markets, high-speed network infrastructure within a building is an intelligent amenity that is a must have, and not a hoped for."*

To all the big property owners and property management firms: Think it cannot happen to your building? In the past, a building in downtown Phoenix which did not have enough spare capacity to lease up the building to 100% occupancy. The property management firm (JMB) found out when it tried to lease out the last 20% of the building and discovered there were no more cables coming into the building to support another tenant.

The area was growing so fast, the phone company diverted the spare capacity from the existing building to service another building which was just being built. The end result was that no new capacity could be installed for another 18-24 months. So the building would have been profitable, but it became a less than break-even building because the amenity was not being managed, let alone understood, by the property manager.

CARLINI-ISM: *"You must understand the problem in order to apply the right solution."*

TELLING THE STORY IS AS IMPORTANT AS MAKING THE STORY

In today's age of the Internet of Things (IoT), if organizations fail to incorporate their communications based information systems strategy as a platform for their overall business strategy- they might as well have no strategy. AND, they won't achieve Best Practices, either.

Depending on who you believe in, the amount of wireless devices is supposed to grow from over 10,000,000,000 today to anywhere from 30,000,000,000 to 75,000,000,000 devices by the year 2020. No one knows for sure how large the growth will be. **(See CHART 7-5)**

CHART 7-5: GROWTH IN THE NUMBER OF WIRELESS DEVICES

YEAR	NUMBER OF WIRELESS DEVICES
2014	10,000,000,000 devices
By 2020	30,000,000,000 devices (ABI Research prediction)
	50,000,000,000 devices (CISCO prediction)
	75,000,000,000 devices (Morgan-Stanley prediction)

Achieving Total Quality Management (TQM), or in today's buzz phrase: "Best Practices" means knowing how to apply technology to buildings. And, "Best Practices" are a moving target to aim at. What was accepted last year may already be obsolete this year. The same issue pertains to "rules-of-thumb" this year. The growth of wireless devices is going to affect how things get built as well as maintained.

In Silicon Valley, you have people working together for a common cause. Venture capitalists, entrepreneurs, bankers, politicians, academia, and most importantly, media make up the "Recipe For Economic Success."

Media? Yes, the secret ingredient. In Silicon Valley, they have people in the media who understand the technology as well as the total picture. And, they can articulate and spotlight it which creates a larger and larger following nationwide, if not internationally. They attract broader, interested sources for venture capital funding as they articulate what is being developed as well as what endeavors need more funding to get to beta testing, prototype development, or to the next step in commercialization of an emerging technology.

(See CHART 7-6)

CHART 7-6: THE RECIPE FOR ECONOMIC SUCCESS

(All of these have to mix and move in the same direction.)

Entrepreneurs	The "catalyst" of the IDEA
Angel Investors, Venture Capitalists	The beginning financing
Engineers, Salespeople	The builders, customizers and sellers of the products
Bankers	The sustaining financing
Politicians	The source of grants and other help
Academia	The source for ideas and future workforce
Real Estate Developers	The builders of the platform for the business
Media	The glue that ties everyone together

In every regional area, if you want to understand if they are maximizing their potential infrastructure, just tune into their local television and radio stations to hear how their news teams and reporters discuss technology and its importance to the region.

If all you hear are fluff pieces and stories which are not in-depth enough to reflect the impact of technology onto the businesses and economy, chances are, there is not anything of significance going on. And if there is, the reporters do not know how to report it, because they simply don't understand it. The perception of investment (or expense) of a project is important. **(See CHART 7-7)**

When this happens, there is no spotlight on applying technology to business. There is no spotlight in adding intelligent amenities to buildings and business campuses.

CHART 7-7: PROJECT PERCEPTION

INVESTMENT VERSUS EXPENSE

Viewed as a Positive	Viewed as a Negative
Asset	Expense
Incorporates payback	No payback perceived, a giveaway
Part of Infrastructure	Throwaway cost
A "must" to expand business	A cost of doing business
Residual value	No residual value

Source: James Carlini © 2012

Throughout the years, many selling technology sold from a *"if it works over there, it should work here for you"* strategy that fell short on many systems and networks having to be customized in order to provide maximum performance.

Cross pollinization of technology is great when you can do it, but in reality, most systems, networks, and applications need specific tweaking and customization for each customer in order to provide a maximum performance.

A "one-size-fits-all" approach has been used by many throughout the years who do not understand what works in one organization may not have any positive results in another. Differences in culture, organizational politics and other variables may affect the success of implementing or adapting to something new.

ADDING INTELLIGENT AMENITIES TO THE MIX

In some of today's more advanced business campuses, there is a different approach to Master Planning by the Master Developer. In these next-generation campuses, power and connectivity are put in during the planning phase and not added as an afterthought after corporate tenants have moved in.

Besides specific intelligent amenities, the concept of "one-stop shopping" and fast-track build-out capability facilitated by a streamlined development process from the municipalities and master developer working together are also concepts that have traction with today's corporate site selection in the United States. This also includes having a Menu of Intelligent Amenities business tenants can select from including a Common Campus Response Team which provides "one-stop supporting." **(See CHARTS 7-8A and 7-8B)**

ECONOMIC DEVELOPMENT EQUALS BROADBAND CONNECTIVITY

Real estate development and its subsequent marketing strategy have to adapt in order to fit the new dynamic demands of prospective tenants dealing within the global economy. In some parts of the United States, there is a multi-level government concern for regional sustainability and job growth. This has created a new unity in community development between municipalities, counties and the developers of next-generation business and industrial parks.

The importance of broadband connectivity (multi-gigabit speeds) has become a baseline amenity requirement and not an option for selecting the corporate "right site." The same goes for redundant power sources connected to the corporate facility. All of this has to be part of the supporting fabric of the underlying infrastructure which supports the facility.

Master planning has to include power and network infrastructure planning as well as some actual implementation before a potential corporate tenant will get serious in considering the location as the "right site" for their facilities.

Redundant power facilities and multiple network carriers providing diverse routing to different networks will attract and maintain a higher caliber of corporate tenants.

These corporate tenants will be a positive addition to the local and regional economy, so the community needs to make sure these employers do not stumble on zoning issues or other typical minutiae which takes a lot of time to unravel.

CHART 7-8A: SAMPLE MENU OF INTELLIGENT AMENITIES

AMENITY	DESCRIPTION	ACCEPTANCE
POWER & TELECOM	ELIMINATION OF SINGLE FAILURE POINTS	BECOMING MORE OF A GIVEN AND NOT A "HOPED FOR" AMENITY
Broadband Connectivity	High bandwidth (1Gbps or more, today 10Gbps, tomorrow 40-100Gbps), multiple carriers, multiple access points	More and more want this. For technology campuses besides speed, diversity in carriers, and connectivity into NLR via Starlight
Power	Pre-planned layouts and capabilities, multiple providers and power grids	This is becoming part of the upfront Master Planning process instead of an afterthought.
Alternative Energy Source	Besides power from another substation or grid, a third-party provider or an on-campus alternative. Peaker Plant or Windmills (Wind turbines generating power)	Diversity of Power source has become a critical issue from several standpoints: Security as well as Compliance issue. Also from a Green Building perspective for renewable energy capabilities.
SERVICES & SUPPORT	DESCRIPTION	ACCEPTANCE
(MASTER PLAN-NING) Upfront Coordination	Quick development process .that includes coordination of all municipal planning issues, adherence to building codes, etc.	New concept but effective in selling to tenants wanting a rapid development process not hindered by municipal issues.
(MASTER PLAN-NING) "Network Tailor" instead of a "Network Jailor"	Custom-tailored network infrastructure. Not an "off-the-rack solution" or a "one-size-fits-all" approach.	A new concept of tailoring bandwidth and broadening choices of carrier services to the demands of tenants.
(CONTINUAL SUPPORT)	Group of diverse service providers including power	A new concept gaining ground as well as popularity. The concept of "one-

Common Campus Response Team Services	and network carriers as well as companies like AT&T, CISCO and other support services	stop shopping" being morphed into "one-stop supporting" after the tenant is in place and is part of the campus.

CHART 7-8B: IMPACT OF INTELLIGENT AMENITIES

AMENITY	ACCEPTANCE	BUSINESS IMPACT
POWER & TELECOM	BECOMING MORE OF A GIVEN AND NOT A "HOPED FOR" AMENITY	ELIMINATION OF SINGLE FAILURE POINTS. CREATES A STRONGER BASE
Broadband Connectivity	More and more want this. For technology campuses besides speed, diversity in carriers, and connectivity into NLR via Starlight	Acceleration of network traffic provides faster access and dissemination of huge documents, drawings, video images.
Power	This is becoming part of the upfront Master Planning process instead of an afterthought.	Power issues are a major concern of businesses today. Eliminating this concern creates higher productive environment.
Alternative Energy Source	Diversity of Power source has become a critical issue from several standpoints: Security as well as Compliance issue. Also from a Green Building perspective for renewable energy capabilities.	Savings on power provides a positive impact on environment and also creates more reliable infrastructure.
SERVICES & SUPPORT	ACCEPTANCE	IMPACT
(MASTER PLANNING) Upfront Coordination	New concept but effective in selling to tenants wanting a rapid development process	Faster ramp-up speed from decision to build to move-in date. Saves time to get productive.

	not hindered by municipal issues.	
(MASTER PLANNING) "Network Tailor" instead of a "Network Jailor"	A new concept of tailoring bandwidth and broadening choices of carrier services to the demands of tenants.	More cost-effective solution and wider selection of services.
(CONTINUAL SUPPORT) Common Campus Response Team Services	A new concept gaining ground as well as popularity. The concept of "one-stop shopping" being morphed into "one-stop supporting" after the tenant is in place.	Faster response to complex problems that impede productivity. Accelerated response equals no loss in global competitiveness due to business support issues.

Source: James Carlini. All Rights Reserved.

How Do You Pay For This

The next big question is, how does all of this get paid for? There are ways to get infrastructure paid for from different sources. You can get federal money, state money, and other grants depending on what improvements you want to make. The three main players in this funding matrix are the owner, the public, and the government.

Depending on how each one of these entities view the project, the way it will get ultimately funded – or not funded, is dependent on the perception each major player has on the project. **(See CHART 7-9)**

CARLINI-ISM: *"We are past the Industrial Age, past the Information Age and into the mobile Internet Age. Let's not reinvest in the* *past, when there is so much to do for the future."*

Those in real estate and property management have to see this paradigm shift in their industry. If they fail to recognize these major changes in what first-class corporate tenants are seeking, they will fail to attract and maintain a first-class tenant mix. This will be detrimental not only to their business, but to the regional economy as well which is always hungry for positive job creation.

The same warning can go to those selling multi-tenant residential units. They also must present intelligent amenities to those wanting capabilities to telecommute or just be able to connect up with their jobs on a casual basis.

CHART 7-9: FUNDING MATRIX FOR PUBLIC INFRASTRUCTURE

The Public's Perspective

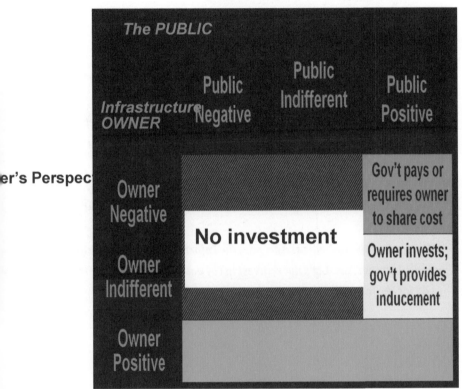

Source: Jerry Brashear, ASME

Owner invests voluntarily – The business case is made

CHAPTER REVIEW

CARLINI-ISMS

"20th Century solutions will not solve 21st Century problems."

"A Smartphone application isn't any good if you cannot access it."

"In today's commercial building markets, high-speed network infrastructure within a building is an intelligent amenity that is a must have, and not a hoped for."

"You must understand the problem in order to apply the right solution."

"We are past the Industrial Age, past the Information Age and into the mobile Internet Age. Let's not reinvest in the past, when there is so much to do for the future."

[8]

NEW PLATFORMS FOR GLOBAL COMPETITION

"Intelligent Business Campuses vs. Intelligent Industrial Parks"

"You cannot build a 21st century regional economy on a weak infrastructure." – JAMES CARLINI

Traditional approaches to building industrial parks and business campuses are giving way to new concepts which include new ways for upfront planning as well as implementations. This emerging trend is evident in commercial real estate developments across the world, not just in the United States.

The concept of Intelligent Business Campuses (IBCs) has come of age as all industries seek out high-performance real estate platforms to launch and maintain state-of-the-art facilities which support their core businesses.

In this case, high performance means high profitability through high occupancy and utilization of redundant power grids and broadband connectivity as well as high security for the total development.

These next-generation business campuses do not have to be situated in urban areas. They could be built in a rural area and the economic impact would be as significant. When you look at some of these next-generation parks, you see how they incorporate intelligent amenities upfront to provide a better *Platform for Commerce* for today's new enterprises.

New Concepts In Developing Next-Generation Real Estate

Before this happened, industrial parks were fairly unsophisticated with the property owner doing minimal amounts of building out the infrastructure and waiting until potential corporate tenants signed up for space.

After they signed up, the property owners would then figure out what they needed to do to finish out the space with network and power capabilities specific to the tenants' needs. These capabilities would be added after the tenant would sign up. Today, these capabilities need to be incorporated upfront.

Thirty years ago, Intelligent Buildings were basically a standalone building. The "intelligent amenities" were focused within the building.

Today, some of those ideas have spread and are now being focused on the campus or business park. Instead of having some intelligent amenities within a building, the whole campus or business park has these capabilities integrated into the fabric of the campus.

In the competitive markets we are in today, more upfront planning and actual implementation of intelligent infrastructure is required to attract and maintain corporate tenants. This is a global phenomenon and it is not singularly tied to the United States and its domestic commercial property development.

Next-generation Intelligent Business Campuses (IBCs) have been built in the United States. In Japan, mainland China, and Taiwan, Intelligent Industrial Parks (IIPs) are being built and utilized in the same fashion. Both IBCs and IIPs are keys to future economic development and regional sustainability for both continents. This is a global phenomenon as governments and commercial enterprises collaborate for accelerating economic development and maintaining regional sustainability.

CARLINI-ISM: *"20th century real estate strategies and solutions will not fit or satisfy 21st century real estate requirements."*

In the United States, campuses like the DuPage Business Center (formerly called the DuPage National Technology Park) in West Chicago, Illinois are definitely an Intelligent Business Campus (IBC). It utilizes multiple network carriers feeding access to broadband connectivity at 40 Gigabits per second (40Gbps) as well as two separate power grids providing power to it.

The intelligent amenities offered from this next-generation business park are the ones necessary to attract and maintain sophisticated corporate tenants. (It will be further discussed in Chapter 13)

CONCEPTS FOR INTELLIGENT PARKS AND CAMPUSES

Some key concepts were gleaned from different professionals within the real estate market as well as from various corporate executives who were looking for quality space at the time the DuPage Business Center (DBC) was being planned and designed. Some of their observations included:

"Parks of the Past" are tied to old infrastructure.

This is a definitive point to make as new parks and campuses are a huge departure from traditionally-designed commercial developments. From a competitive standpoint, you can dismiss all your competition if they marketing a traditionally designed and built industrial park. Traditionally-designed parks equate to single connectivity to one network carrier as well as singular power sources. That may have been adequate in the "horse-and-buggy days", but not in today's or tomorrow's real estate markets. Any corporate tenant which has mission critical applications, cannot utilize the antiquated amenities of a business park which does not offer redundancy in both power sources as well as network communications infrastructure. (You cannot have a single-point-of-failure. Most buildings do not provide redundancy in either type connection to power or communication carriers.)

Must Develop a Vision for the Park. Don't Build Junk.

This comment was from a physicist from Fermi Labs looking at high-tech real estate development. His concern was when you build something, build quality. Also, he talked about the need to establish some type of theme for the Park.

A theme creates a sense of cohesiveness across the tenants. They are all focused on a similar core business or industry. Themed parks appear to be more successful.

Is There a Theme to the Park? Adhere to it.

The most successful intelligent business campuses (or parks) are those that have some type of common theme across the tenants occupying the buildings. The concept of what gets put in first, dictates what comes in next, is very important as to understanding how many companies select their locations. Synergies need to be recognized, established, and then promoted to potential corporate tenants.

This holds true for parks built internationally as well. Examples in Taiwan and mainland China adhere to this "single theme" approach for their Intelligent Industrial Parks (IIPs) including Far Glory Park in Taiwan, Sunward Park in Hunan province, and Cyberport in Hong Kong, just to mention a few.

CARLINI-ISM: *"What is put in first, dictates what is put in next when leasing up next-generation Intelligent Business Centers (IBCs). You should adhere to a theme."*

The Park has to "Make a Statement" (Theme).

How do you differentiate your property from your competition? This is something relatively new, but substantially important. Are you building a campus or business park for a specific industry? Will the park/campus offer all the buildings addressing different aspects of the manufacturing and distribution process?

For example, if Boeing was to look for a new corporate R&D facility, would they want the facilities close to the manufacturing facility as well as a runway for testing out prototypes? If an Intelligent Business Campus were to support that endeavor, its theme would be some type of Aviation Park and would be close to some type of airfield.

Some ideas for next-generation parks include the following:

- Adaptive Vision – create and recruit a Science Advisory Board.
- Each park should project a theme. (Single vision – like a Logistics Park, Semi-conductor products, construction products, or an R&D park)
- Unity in community development. (Multi-level government concern for regional sustainability. Cooperation in order to accelerate the project's completion, rather than, blocking it with bureaucratic zoning and taxing issues)
- Streamlined Development Process for tenants moving in. (Must be coordinated internally and with different government agencies)
- A totally "Green" park. This could generate so much coverage in the different trade press that advertising it would not be necessary.

This is very critical as the sense-of-urgency to respond to a potential customer (tenant) is paramount. The worst things which can stall landing a potential corporate tenant are hidden fees and lagging timetables when it comes to negotiating with local and county government agencies. Building permits, impact fees, and any other municipal or regional fees must be concise and defined with exact timetables for completion. No one wants to hear, *"We're waiting on approval on these details'*, when it comes to building a new corporate facility.

With the way regional economic development is today, those municipalities and other government agencies who do not establish concise fees and short timetables for permits run the risk of losing good corporate tenants who create jobs.

The potential corporate entity will quickly focus on selecting another regional area to avoid delays, time-consuming red tape, hidden taxes, and other bureaucratic obstacles.

Unity in Community Development

This concept ties in with the above process. You must coordinate the ongoing processes of internal as well as governmental agencies on all issues regarding traffic, infrastructure and overall regional sustainability. Failure to do this will negate any regional economic development and viability in today's competitive marketplace. If all the private and municipal agencies do not have their act together as a region, some other region will. Job erosion is a real issue which most states do not have a good grasp of. The states who do are the ones enticing businesses to locate there and streamlining their bureaucratic process.

CARLINI-ISM: *"When states do not address the right issues, job erosion occurs and they lose their tax base."*

The Ability for Adaptability for the Future to Stay Fresh. (Park must offer flexibility.)

What was considered "state-of-the-art" five years ago, is obsolete today. Many existing properties need more than just a cosmetic facelift. Some need a complete overhaul on their intelligent amenities: network infrastructure and the power grid they connect into.

A single connection to a power source for electricity or for communications, will impact the viability of the campus. Redundancy of power sources coming off of separate power grids as well as multiple network carriers providing broadband connectivity via fiber optics to all buildings within the park (or campus) should be offered. They are a must-have for any corporate enterprise which has mission critical applications.

Another big renovation is adding broadband wireless connectivity and Smartphone support in the form of capabilities like WiFi and DAS (Distributed Antennae Systems). Although this is in its beginning stages, this amenity will quickly become more important as the Internet of Things (IoT) concept becomes implemented. It will also become important to property managers and all municipalities who are concerned about future economic growth and survivability.

CARLINI-ISM: *"Municipalities which create a lot of red-tape and development fees for developers build more walls and obstacles. These obstacles kill economic development instead of streamlining processes which open up doors and invite local economic development."*

Summarizing some of the ideas for next-generation parks include the following:

- Each park should have a theme (single vision – like a Logistics Park or an R&D park or a park focused on medicine or law).
- The more exact the theme, the more successful the park.

- Unity in community development. (Multi-level government concern for regional sustainability)
- Each park should make sure that all of its conduits for both network and power distribution are in place, so they don't have to dig up landscaping or roads after they have been built. Conduits for fiber and electrical should be put in upfront, and not as an afterthought.

Infrastructure which includes intelligent amenities like power and connectivity must be reviewed at each park/campus to assess their capabilities. Do they address what today's and tomorrow's corporate needs are? If a park or campus has one network carrier supporting it and only one source for power, the park is going to fall into a secondary market position.

No matter what the owner tries to market, the bottom line is the park's intelligent amenities do not address what prime corporate tenants are looking for today and for tomorrow. (They are looking for redundancy when it comes to power and connectivity.)

CARLINI-ISM: *"Mission critical applications require redundant power sources as well as redundant network services. Redundancy is NOT an option."*

Ever-changing commitment to tenants.
This concept ties in with the concept above. Every industry is evolving and in order to maximize returns, property managers and owners must insure tenants are being taken care of in various areas of property management. Adding new capabilities can increase the viability of the park as well as the surrounding economic region.

World-class Park (Must offer everything to go with it. World-class Advisory Board?)

Some sophisticated parks with administrative boards focus on building and sustaining a solid position for the park within the community as well as region. This is very important to gain all these perspectives and have them contribute to the fostering of good governance over the park. The better the direction a park establishes, the more sustainable economic development will come to the region.

Shared Facilities - International Conference room, Shared back-up generators, etc.

What do you offer in the park or business campus? Are the amenities unique? Are they marketable to a specific target market? Can they be sold as a unique amenity against other properties within the area? This is important from a point of initial marketing as well as ongoing marketing. This is an area requiring creativity and a good knowledge of the existing market, the potential market, and the tenant mix.

If this is done right, you can eliminate about 90% of the competition. With that type of elimination, no need to drop prices on leases (or land purchases if you are selling off parcels within the campus).

Turn-key infrastructure

This is a relatively new approach to Intelligent Business Campuses and Intelligent Industrial Parks. It has become a competitive necessity. Companies considering moving in, want to see infrastructure in place upfront before they make a decision as to what property they want to build on. Intelligent amenities, such as network connectivity and power capabilities, which were considered an afterthought to the property developer, are now regarded as a must-have upfront and not a hoped-for.

CARLINI-ISM: *"Master planning today is not only a new ball game, it's a new sport."*

Adding velocity (acceleration) to the transference of technology from research to commercialization.

By having all the manufacturing and R&D elements together, getting a product from the initial prototype to commercialized production is a shorter time period. **(See CHART 8-1)**

The importance of this goes beyond the Park and the local region. By accelerating the development process, getting new products out to the market can add to the national economy as well.

CHART 8-1: THE STEPS TO TRANFERENCE OF TECHNOLOGY

R&D ⇨ **Manufacturing** ⇨ **Distribution** ⇨ **Retail** ⇨ **Customers** --- *everything towards complete technology transference to the end customer.*

Menu of Intelligent Amenities (Park & Buildings)

What do you offer? Dual power feeds to two separate power grids? Maybe an alternative power source is offered (secondary power company, wind turbine, solar panels)?

Broadband connectivity? How many broadband carriers feed into the park? (two are good, three are better, more than three – great)

These are becoming necessary amenities which are being more and more demanded.

A detailed list for corporate site selection committees as well as prospective tenants should be generated and kept up-to-date. This is a good way to separate out the better properties from the ones which don't offer what is needed today.

It is nice to eliminate 90% of your competition. Once you can do that, you don't need to drop your square footage pricing like so many commercial property groups do as their "last resort" strategy.

CARLINI-ISM: *"Any intelligent amenity eliminating 90% of the competition should be seriously considered a top priority for any real estate developer or owner."*

Must have graphics to show comparisons.

A good visual yardstick backed up by features and capabilities to compare competitive properties and campuses is a must. If you are not providing the yardstick to measure significant amenities and alternative solutions – your competition will be.

This is a proven approach I used in both the commercial real estate and the network carrier/regional Bell operating company industries.

CARLINI-ISM: *"You must develop metrics and comparisons for high-tech real estate. If you don't, you will be measured by your competitor's yardstick and you will always come up short."*

POTENTIAL CONCERNS WITH PARKS AND CAMPUSES

Close to an Airport

Although other next-generation business parks like Far Glory in Taiwan and DuPage Business Center (formerly the DuPage National Technology Park) in Illinois are close to airports, they can be viewed in a positive light as well. In the past, being too close to an airport was a common concern about business parks and viewed negatively, but that is slowly changing.

Power grids

Must check to make sure dual feeds or alternative power sources like wind turbines, secondary utility, solar panels are available. Those properties demonstrating the capability of providing power from two or more alternative sources should be deemed to be much more viable for 21st century real estate demands than those which cannot.

Lease only

Can or cannot buy land? This is a big issue with some corporate tenants. They want to be able to purchase the land. There is a need to correctly market to potential tenants if it is a lease-only arrangement for the campus or park. Some deals will fall through if the option to purchase the land to build on is not part of the offering.

Create the first "Green Park." (This would be a big payback) Detention Ponds, Bike Path, and other "green initiatives."

What approaches are being taken to conserve energy and the environment? This has become a large area of concern with more talk than actual implementations. This will become a larger issue as time goes on and "green capabilities" will be a part of the amenities list.

If you create an "All-Green Park focused on "Green" issues. If you put in an extra $500,000 for Green amenities, but you receive $2,000,000 in free press to acknowledge those refinements, you are ahead. The extra spotlighting will attract prospective tenants.

"Create a Pond" - retention pond, reservoir, water conservation, and other "green" elements in an exhaustive list to show prospective tenants as well as the press.

Green Building Requirement (for existing parks)

Could your existing park turn into a Green Park? What impact would that have on advertising, corporate sponsorship, name recognition and other significance? By going "green", it can translate into free advertising by media who want to highlight these improvements and environmentally-friendly amenities. Again, for every dollar spent, you could get back two or three in advertising and promotional consideration. This is something to consider when weighing the extra costs of "going Green" in the development process.

Renewable Energy

MUST be in the fabric of the park's infrastructure. If not, you shut off possibilities for positive recognition.

KEY POINTS AND CONCEPTS TO FOCUS ON

Fast-track build-out capability.

This is a necessary zoning/ building code component for local government to provide to someone who is trying to develop a large park.

There must be coordination with local municipalities and their building permits, utilities, and impact fees in order to package a "fast-track offering" to potential tenants.

This is a must-have and not a hoped-for in today's development process for business parks and industrial campuses. **(See CHART 8-2)**

CHART 8-2: PARK/CAMPUS APPROACHES

WITHIN THE PARK/ CAMPUS	MUNICIPAL APPROACH
BUILDING AMENITIES	EMPHASIZE QUICK APPROVAL PROCESS
BUILDING OPERATIONS	EMPHASIZE QUICK APPROVAL PROCESS
PARK/ CAMPUS AMENITIES	EMPHASIZE QUICK APPROVAL PROCESS
PARK/ CAMPUS OPERATIONS	EMPHASIZE QUICK APPROVAL PROCESS
UTILITIES	EMPHASIZE QUICK APPROVAL PROCESS
OUTSIDE OF THE PARK/ CAMPUS	COMMUNITY & SUR-ROUNDING COMMUNITIES
ZONING ISSUES	HAVE A UNIFORM PROCESS
BUILDING PERMITS	HAVE A UNIFORM PROCESS

TRAFFIC/ ROAD APPROVALS	HAVE A PRE-AGREED UPON PROCESS
INTERGOVERNMENTAL ISSUES	HAVE A PRE-AGREED UPON PROCESS
INTERGOVERNMENTAL AGREEMENTS	UNDERSTAND WHAT IS NEEDED AND GET THESE APPROVED BEFORE DEVELOPING THE SITE

Upfront infrastructure (Fairly unique. Could this become a requirement?)

What you build out first will attract or repel the next potential tenants. You really need to figure out what amenities will be provided and what will be a tenant add-on.

Rapid Response Team of connectivity services (AT&T, CISCO, etc.)

STREAMLINED DEVELOPMENT PROCESS

1 - Adding velocity (acceleration) to the transference of technology from research to commercialization to get new products from prototype to commercialized product in a shorter period of time.

2 – Next-generation Network Access Point ((this could be access to a carrier hotel on the Park's premise where the tenant can choose from several national carriers as well as specialized research networks like NLR (National Lambda Rail)).

3 - Look for theme or signature for Park. No common set of success factors or amenities. All parks are different.

THE GLOBAL MARKETS ALREADY EMPLOY IBCS AND IIPS

We need to increase the amount of next-generation industrial/business campuses being developed in order to become more competitive and keep up with global initiatives which are being pursued in various international markets.

RE-TOOLING COMMERCIAL REAL ESTATE CURRICULA

Hear that huge shattering sound of multiple stories of glass crashing down to the ground? It is the shattering of decades' worth of commercial real estate concepts which don't work anymore.

The sad part is many in the Commercial Real Estate (CRE) market need to tune into this deafening thunder. The status quo has been shattered and shattered again. As I said in an earlier article: (HTTP://ONPURPOSEMAGAZINE.COM/2013/02/20/DEFINING-21ST-CENTURY-REAL-ESTATE-2/)

Selling products and services into this 21st century integrated real estate environment requires an expertise in understanding multi-disciplinary skills and next-generation solutions. It is a multi-level sell when trying to promote new intelligent amenities for next-generation buildings, multi-venue entertainment centers, and intelligent business campuses.

The new real estate concepts are not new, they are just not being adopted as quickly as they should be by CRE professionals. Want to sell high-tech real estate into today's sophisticated market? These are the people you need to sell to: **(See CHART 8-3)**

CHART 8-3: THE MULTI-LEVEL SELL

The Building Owner (Property Owner)
The Developer
The Financier
The Property Manager
The Leasing Agent
The Tenant(s)
The Media *

* Yes the media, because they can help sell the whole package or they can kill it by not presenting its unique qualities because they don't understand it.

Examine the graduate school curriculum at most major universities in the United States. Real estate curricula are at best, based in the 1950s and at worst, in the 1850s. Is that an exaggeration?

They don't even offer one course on Intelligent Buildings, let alone a concentration in them (for either a major or a certificate program). Intelligent Buildings have been around for 30 years.

Except for some universities in Asia, the University of Reading in Great Britain and one in Dubai, most schools teaching real estate have not added degree programs or even single courses which should be offered to train today's graduates on tomorrow's building amenities.

Sometimes Intelligent Building strategies are mutually exclusive. Before real estate companies can set any strategy, everyone should agree on what the organization's direction is.

How can you set a direction if you don't know the advanced concepts, let alone the advanced principles of intelligent real estate? How can you develop the right mix of buildings and amenities for a target market? How can you sell prospective corporate tenants?

Some schools might argue they offer a "Special Topics" course which includes "emerging trends" but Intelligent Buildings should have had their own course by now after three decades, if not a whole section of courses dedicated to these new concepts.

Review of United States initiatives and intelligent business campus development, IBCs versus Asian initiatives in Hong Kong, Taiwan, and Mainland China's Intelligent Industrial Parks (IIPs) could in itself be a potential course.

WHAT SHOULD BE IN A CURRICULUM?

There should already be courses on:
- Intelligent Building Concepts (A general overview to get familiar with terminology as well as basic applied principles),
- Intelligent Infrastructure (the **Platform for Commerce** which focuses on supporting regional economic development), and
- Courses on Next-Generation Real Estate:
- Intelligent Business Campuses (IBCs) (The next-generation business campus, industrial park or technical campus. IIP is the acronym used in Asia for an IBC.)
- Intelligent Infrastructure (II) (Supportive infrastructure like power grids and broadband connectivity)
- Intelligent Retail/ Entertainment/ Convention Center complexes (IRECs) (Study of multi-venue campuses of retail stores, entertainment/restaurant centers and convention and/or sports arena built to provide new Smartphone

applications that cross-market the products and services of those businesses within the "umbrella" of the DAS (Distributed Antennae System) wireless network.)

Different spheres of disciplines are starting to intersect and overlap with each other. The dynamics of most schools are to teach a single discipline and a single focus. Today, four diverse spheres are intersecting at critical junctures, curricula need courses which discuss the combined concepts of these new converged principles and how they impact the market.

The more graduates who have a grasp of the new multi-disciplinary focus, the more effective their ongoing leadership and team management will become as the real estate industry evolves. This also applies to those in regional economic development.

They need to understand the building blocks for today's and tomorrow's economies, not those that built yesterday's economy.

CHAPTER REVIEW

CARLINI-ISMS

"20th century real estate strategies and solutions will not fit or satisfy 21st century real estate requirements."

"What is put in first, dictates what is put in next when leasing up next-generation Intelligent Business Centers (IBCs). You should adhere to a theme."

"When states do not address the right issues, job erosion occurs and they lose their tax base."

"Municipalities which create a lot of red-tape and development fees for developers build more walls and obstacles. These obstacles kill economic development instead of streamlining processes which open up doors and invite local economic development."

"Mission critical applications require redundant power sources as well as redundant network services. Redundancy is NOT an option."

"Master planning today is not only a new ball game. It's a new sport."

"Any intelligent amenity eliminating 90% of the competition should be seriously considered a top priority for any real estate developer or owner."

"You must develop metrics and comparisons for high-tech real estate. If you don't, you will be measured your competitor's yardstick and you will always come up short."

[9]

AIMING FOR QUALITY IN TECHNOLOGY INVESTMENTS

"In the long run, men hit only what they aim at. Therefore, they had better aim at something high." – **HENRY DAVID THOREAU**

I n any real estate development or core business enterprise, there is a basic need to procure goods and services. It is important money is not wasted and the critical management strategy is to aim for quality in any investment. This also relates to aiming for quality when applying technology to the organization as well.

Whether your organization is a large Multi-National corporation or a small partnership, one of the common problems that face all businesses is selecting the right strategy to procure and manage the applications of technology to the organization.

There are some Captains of organizational Titanics out there still waiting for the right person to come along and re-direct their organization's communications and information systems infrastructure into a starship built with strategic advantages. Others do not understand where they are headed, nor have the skill sets which are needed to navigate and establish the right course for action for this global market in this century. Here is my navigational aid for applying technology to the organization.

Years ago, I developed the **TARGET** Map of Technology. **TARGET** stands for "Technology And Revolutionary Gadgets Eventually Timeout." Understanding this concept is a must for any organization utilizing technology.
(See CHART 9-1 below)

When the stimulus packages came out to try to revitalize the US economy back in 2010, it focused more on the first three layers of the *Platform for Commerce*. Not much money was ear-marked for other infrastructure levels.

The **TARGET** Map's Five Levels of Stages of Technology include: **Embryonic (1), Proven (2), Accepted (3), Safe (4), and Obsolete (5)** layers.

This well-tested conceptual approach was used to explain emerging technologies which over time lose their competitive advantage, become competitive necessities, and eventually become competitive disadvantages to maintain. (It was also presented in a white paper entitled, *"Aiming For Quality in Technology Investments"*, published by the International Engineering Consortium in their Annual Review of Communications, 1996.)

CHART 9-1: THE FIVE LEVELS OF TECHNOLOGY

Technology And Revolutionary Gadgets that Eventually Timeout

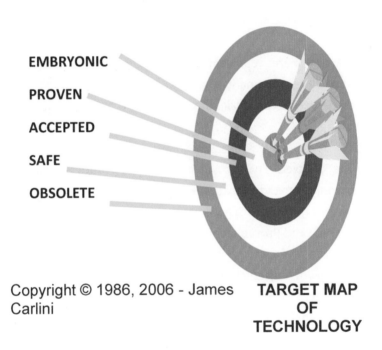

EMBRYONIC

PROVEN

ACCEPTED

SAFE

OBSOLETE

Copyright © 1986, 2006 - James Carlini

TARGET MAP OF TECHNOLOGY

Over the decades, I was brought in to review many multi-million dollar projects to insure money being spent on technology was well-spent and as Steve Duerkop, the Senior Group Vice President of JMB Property Management, once said, *"We want to get $20,000,000 worth of equipment and services for the $20,000,000 that we are laying out for this building renovation."*

No matter what technology you are assessing, it eventually passes through every layer of the **TARGET** Map of Technology. (**TARGET** stands for Technology And Revolutionary Gadgets Eventually Timeout.)

The **TARGET** map is a template which can be overlaid across any organization to see exactly where money and resources are being spent. In order to become more competitive, resources should be expended more towards the center of the **TARGET**.

The center area represents more risk (but there is also more reward because a larger competitive advantage is created) and as you work out to the outer circles, risk diminishes, but so does the competitive advantage.

The "Obsolete" layer provides no competitive advantage and in fact, adds competitive disadvantage because you are paying to maintain obsolete capabilities. Those resources would be better spent on Embryonic and Proven technologies. **(See CHART 9-2)**

The prime objective on any real estate development or property management technology endeavor is to get your money's worth. Adding technology, as an intelligent amenity underpinning, means understanding the procurement process for evaluating and purchasing complex systems and services.

A **DART** is a Designated Active Resource with Timeframes. Every organization has a limited amount of **DART**s to apply to the **TARGET**. The primary question becomes, are you hitting the center (competitive advantage) or wasting resources in the outer rings (competitive disadvantage)?

Planning is as easy as throwing "**DARTs**" onto the **TARGET** map. This model has been used as part of discussions in both undergrad and Executive Masters programs at Northwestern University focusing on planning and managing technology within an enterprise.

It is a good visual to help understand where technology dollars are being spent in the enterprise and where they are being wasted. It helps answer the question, *"Where should we spend our resources?"*

CARLINI-ISM: *"You must understand the usefulness of the technology infrastructure servicing your enterprise."*

How can we look for residual value in developments and applying technology to the enterprise? How do we prioritize projects and new implementations? These questions need to be asked and answered.

EXPLORING THE LEVELS OF TECHNOLOGY

Let's segregate the different levels of technology an organization has to assess, select, implement, and manage.

We could group the different levels of technology into five levels of TARGETs on an imaginary dartboard. **(See Chart 9-1)** These levels also represent the amount of risk and the competitive advantage that will be assumed for each technology and service. Let's define the diagram as the **TARGET** map of the enterprise and the DARTs as our organization's Designated (or Dedicated) Active Resources and Timeframes.

CHART 9-2: COMPETITIVE IMPACT OF TECHNOLOGY

LAYER	DESCRIPTION OF TECHNOLOGY	COMPETITIVE IMPACT
1	EMBRYONIC *(R&D – NEW IN-NOVATION, HIGH RETURN FOR IN-VESTMENT)*	HIGHEST RISK, HIGHEST COMPETITIVE ADVANTAGE. *(IF SUCCESSFUL)*
2	PROVEN *(BETTER RETURN FOR INVESTMENT)*	RISK. COMPETITIVE ADVANTAGE.
3	ACCEPTED *(GOOD RETURN FOR INVESTMENT)*	LESS RISK. COMPETITIVE NECESSITY.
4	SAFE *(FAIR RETURN FOR INVESTMENT)*	NO RISK, BUT, ALSO NO COMPETITIVE ADVANTAGE.
5	OBSOLETE *(WASTING MONEY)*	RISK BECAUSE COMPETITITIVE DISADVANTAGE TO MAINTAIN THIS LEVEL.

If your organization can aim at those high risk technologies in the **TARGET** MAP by dedicating active resources and timeframes (**DARTs**) to develop them, it will be well on its way to gaining on, and out distancing, its competitors. **DARTs** are limited. **(See in CHART 9-3)**

One limitation is time. You cannot run a machine or application longer than 24 hours a day and with people, they cannot work 24/7 unless you have three shifts of people performing the same tasks.

CHART 9-3: TYPES OF DESIGNATED ACTIVE RESOURCES W/ TIMEFRAMES (DARTs)

DESIGNATED (or DEDI-CATED) ACTIVE RE-SOURCE	TIMEFRAMES
PEOPLE	*40 HOURS A WEEK (+)*
HARDWARE	*24/7*
SOFTWARE	*24/7*
SYSTEMS	*24/7*
AUTOMATED MACHINES	*24/7*
DIAGNOSTIC EQUIPMENT	*24/7*

The first level of the **TARGET** MAP, the innermost circle, would encompass the newest and emerging technologies which are at an embryonic state. These technologies have not been thoroughly tested, or fully implemented, in any real environment. At best, they would be categorized in the prototype stage. This level, **EMBRYONIC** technologies, represents the highest level of risk, but also the shortest route to competitive advantage through applying technology based solutions. An example might be a multi-terabit router or multiplexer.

The second level, **PROVEN** technologies, encompasses those technologies which have been proven in a real working environment and are now being marketed by more than one firm as well as being utilized in more than one industry application. Examples of this level could be fiber optic communications (SONET), a sophisticated Smartphone application or optical disc storage devices. This level still has a substantial element of risk, but will also provide an organization with a significant advantage over most competitors.

This second level is where most executives believe to be the starting point where their organizations' systems should begin. No one likes to be a pioneer with implementing new technology on a test site basis and some believe that it is risky enough to have applied technology at this second level.

The third level, **ACCEPTED** technologies, would encompass those technologies that have been proven to a point where they were readily accepted by many and the need to provide formal training and education to general users would be present. Examples of this level would include specialized tablet applications, video applications, and satellite communications. At this point, risk is minimal but impact on an organization is viewed as just keeping up with everyone else - a traditional, conservative approach.

The fourth level, **SAFE** technologies, would include all those products and services that the general public would use on a regular basis. Examples of this level would include cellular phones, cable TV, iPODs, and other consumer products.

At this level, the technology is in such a mature state it provides no real impact to an organization looking for a technology based edge over their competition. If anything, technologies at this level are competitive necessities. You have to have them in order to compete.

The fifth level, **OBSOLETE** technologies, would be antiquated and obsolete technologies. These would have nothing more than historical significance with little current value or application. Continuing to use these technologies can be hazardous to the organization's health within their marketplace.

Examples of **OBSOLETE** technologies would be vacuum tube technology, mechanical adding machines, punch-card machines, or paper-tape terminals. Some would also deem DOS-based applications as obsolete, yet I know of several financial institutions utilizing this software to maintain certain financial applications. At this point, industries and individual organizations relying on this level of technology would be prime candidates for significant immediate change or bankruptcy.

Within an industry that has not automated or has not applied many new technologies, the organization which first implements a new strategy employing technology as a competitive tool creates a whole new playing field for the competition to adapt to. While the competition is catching up, the pioneering organization is enhancing their new competitive edge and may even be considering moving to a second generation or second application in order to sustain their competitive advantage.

If this implementation continues, the first one or two organizations will have pulled far ahead of their competition and may have even acquired some of the less efficient players. They sustain their advantage and lead over the competition.

Sustainability of the organization's competitive advantage gotten from technological advances is critical, according to Charles Wiseman, former professor at Columbia University's Graduate School of Business and Competitive Applications. His observations followed Michael Porter's, who believed that competitive advantages must be sustainable in order for an organization to stay ahead of its competition.

CARLINI-ISM: *"Applying network based information systems to sustain a competitive lead is not a one shot deal. It's a continuous process."*

In a competitive industry which has a mature understanding of the use of communications based information systems, an **EMBRYONIC** level of technology might be needed in order to create a competitive advantage over the other organizations. A good example of this would be the brokerage industry or the banking industry where the applications of technology take place rapidly and are clearly viewed as sources for competitive advantages.

In a slower paced marketplace, such as the frozen food industry, implementing a **PROVEN** level or even an **ACCEPTED** level of technology might give a significant competitive advantage over the rest of the market players. It would also be perceived as having as much risk as implementing the higher level technology in the industry that is used to more technology application endeavors.

Depending on the industry, a technology might be categorized at a different level. A level two technology (**PROVEN**) in one industry may have already dropped to a level three (**ACCEPTED**) technology in another industry. Also, the rate of progression through each level of technology will be different for each product or service depending on their availability, acceptance, and application. **(See CHART 9-4)**

CHART 9-4: TECHNOLOGY IS DEPENDENT ON:

1	AVAILABILITY
2	ACCEPTANCE
3	APPLICATION

Because every organization has a limited number of DARTs to work with, it is important for executives to realize that an investment in a riskier technology may pay off for a longer period of time than one that is well on its way to becoming obsolete.

NAVIGATING THROUGH TECHNOLOGY

Let's discuss a hypothetical case of an organization needing a competitive edge that is willing to be a "pioneer" or assume risk on developing a new application.

We begin with the customer going to a vendor for a specific need. The vendor's marketing team discusses the need with the customer, presents the project to the manufacturing organization which goes to its research and development organization to develop a prototype solution for the customer. Research develops a prototype and the vendor presents it to the customer. The customer tests it and gives feedback to the vendor who, in turn, reviews the information with their research department.

The research organization may rework the prototype, enhancing it with the customers' suggestions, and return it to the manufacturing and marketing organizations. The manufacturing organization creates the product and the marketing team delivers the product.

If everything has gone well, the customer organization now has a definite advantage over their competitors. The competitors may seek out the same vendor or similar vendors to develop a solution to their needs. This pushes the technology into the second level where multiple versions of the technology become available to multiple users in multiple industries. Risk is reduced for those implementing the technology.

As more vendors develop and more customers buy into the technology, the need to address education becomes important. Thus, the third level of technology (**ACCEPTANCE**) is reached. If the technology finds a widespread application, the need to educate many people is critical and formal education courses as well as training seminars begin to appear.

As the technology matures, it may find its way into society as an everyday product or service. At this point, it crosses into the fourth level **SAFE**. From a competitive advantage standpoint, technologies rated in this category are not too useful to the organization trying to keep up with the competition.

All technologies eventually reach the fifth level which is obsolescence. At this point, the organization should have been trying to update the technology and phase out the old technology. In many cases, the use of a technology cannot be stretched into the fifth level. Competition dictates organizations must upgrade their systems long before they reach obsolescence. The movement of technology through each level may be different depending on how fast it is assimilated into the mainstream.

MEASURING TECHNOLOGY

Depending on the industry, the level of technology needed to assume a competitive advantage may reside in a relatively low risk, LEVEL 3 (**ACCEPTED**) technology. In some industries, getting a competitive advantage does not necessarily mean implementing a LEVEL 1 (**EMBRYONIC**) technology.

It is up to the organization and its CIO to classify applicable technologies and make decisions on what should be implemented. Designating a technology as being at a certain level is most accurate if the organization does it in an assessment of what they have installed and what they are looking at implementing. In other words, someone's LEVEL 2 (**PROVEN**) technology might be someone else's LEVEL 3 (**ACCEPTED**) technology.

CARLINI-ISM: *"Setting the standard is the sign of an industry leader. Playing catch up or succumbing to providing mediocrity is the sign of a member of the trailing pack."*

WHAT IS THE RECIPE FOR SUCCESS?

Everyone wants to know the right steps to take to be successful. It's not as easy as some people want to make it out to be. CEOs and their staffs have to determine the right mix of technologies which will do the organization the most good with the least amount of risk or with a pre-determined threshold of risk the organization agrees it needs to assume for competitive reasons. All organizations have a limited amount of **DARTs** to throw at the **TARGET** MAP.

The Chief Information Officer (CIO) and/or the Chief Technology Officer (CTO) should be able to define and prioritize communications-based information technologies which can be used as vehicles for providing new products and services to the customers of the organization.

Once this is accomplished, the information executive must convey the use of technology as a means of providing a competitive edge or cost effective alternative to existing systems and services. The CIO/CTO must be able to sell this pro-active approach of applying technology and creating new business opportunities to senior management. The CIO must act as an organizational navigator that selects the right paths of technology to explore, utilize, and build into highways to success.

As long as it is cost effective or provides a marketing edge, a new technology may find its way into an organization fairly quickly. Today, those organizations going through this exercise are benefiting from several aspects:

-   For one, they are exploring their needs for technology as well as their capacity to take on risk to attain their objectives.
-   The second benefit is that they have invested in creating the internal structure to strategically plan, assess and direct technologies as an ongoing high level management function rather than an erratic, secondary management responsibility.
-   The third benefit is that they are spending some time today exploring and experimenting before they are put into a situation where they have no time to plan or analyze a technology based strategic direction.

WHAT DOES IT TAKE?

Today, it is not good enough to be a caretaker of in place information systems or a cost cutter of network services. This type of approach does nothing for the organization in the long term, but is still used by some executives who want to make a quick fix to impress management. The improvement, if any, is short term and should not be incented with any executive bonuses.

Cost-cutting is not a strategic initiative, it is a tactical function that can and should be accomplished by a low-level manager or analyst.

CARLINI-ISM: *"If an organization is stagnant, all it wants is someone to maintain the current systems – in effect, a network janitor."*

In some cases, being a caretaker means being laid off. In the next decades, organizations will need decision makers and risk takers, not people who "go with the flow" and graze at their desks. Years ago I made the observation that there are two types of people in this industry - cattle and catalysts. The corporate cattle graze at their desk and are indecisive, whereas the catalysts are actively pursuing endeavors and contributing.

Executives in charge of managing new resources cannot take a distant approach to managing the direction of technology. Some people talk about the business school graduate walking in and taking over the top position becoming the Chief Information Officer. This simplistic approach lacks common sense.

There are too many variables that the graduate has to become familiar with before taking on the responsibilities associated with this top position.

The only way that this can come about is if that graduate puts in a solid apprenticeship in managing different technologies from a hands-on basis over a period of time.

Getting a manager who was in charge of the cafeteria or the motor pool doesn't mean that they can manage technology either. Some companies which are viewed as being progressive, adhere to the myth that if you managed in one department, you can manage in any department.

This does not hold true with spinning those managers into an information or communications technology management position. It just does not work. Period. Just ask any of their subordinates. I have seen this and it is a disastrous management approach.

EDUCATION HAS TO CATCH UP

We need to re train managers as well as workers in learning how to work with new technologies. Courses must be taught not only on different skills but also on the process of learning how to learn. In the past, companies such as AT&T, IBM, Motorola, and the Regional Bell Operating Companies (through Bellcore) had extensive training courses to keep their workforce up to date. It is a great management philosophy which should be retained and adopted by others.

CARLINI-ISM: *"A well-trained workforce is a good hedge against the competition."*

For the most part, universities and their curricula are not keeping up with the demands of a more technologically based corporate organization. The classes which are offered do not address some of the newer managerial issues, nor inter-related infrastructure. Closer ties between the universities and the corporations are being formed in order to create a better match of academic skills with business needs along with technology transfer between both.

Several PhDs in Computer Science and Telecommunications were signing up for undergraduate courses in Telecommunications and Local Area Networks at Northwestern University because they felt they did not get a practical perspective on current business needs from the universities they attended for their doctorate. They were frustrated because the courses they had taken did not prepare them for real world issues.

Some academics do not understand we have gone far beyond the need for only programmers, engineers, and other specific skilled people. A person can get a degree in electrical engineering, computer science or business but cannot easily find a curriculum offering integrated technology management or consultative selling and marketing for high technology equipment. Many courses offer specific skills, but do not cover the overall understanding of systems integration and a broad, multi-disciplinary approach to solving business and technology problems.

What is needed are new graduate curricula broadening the perspectives and managerial views of a technically degreed person into strategic applications and integration of technology to business needs.

Senior system integrators parallel them with needing to understand many areas   business as well as technical, being able to assimilate into new technologies in order to provide rapid applications development.

Because education falls into the third level of the TARGET MAP diagram, the only way formal education can help with the strategic management of technology is if it focuses on what is happening in the first two levels of technology and gears part of its curriculum to those embryonic areas which, in some cases, it has   with various degrees of success.

STRATEGIC TRIAD OF TECHNOLOGY MANAGEMENT

Today's executives have to take a more active role in understanding and knowing how to apply technology within their sphere of influence, regardless of the industry.

This requirement leads to the first cornerstone of the Strategic Triad of Technology Management:

YOU MUST KEEP ABREAST OF TECHNOLOGY

It is impossible for decision-makers to direct or manage complex assets in an organization unless they understand the broad impact of technology applied to their business. They must keep up with current developments and philosophies across their own organization's industry as well as related technology based and regulated industries that supply telecommunication based information systems and services.

How do you keep up with technology? With all of the new developments, this is a question some have given up on. Serious executives and organizational decision-makers have made it a point to at least be aware of new developments by reading trade journals, attending seminars, and even going back to school.

In the past, some took a hands-on approach to learn how to use a PC in order to manage their organization by using various software packages which provided management, word processing, and graphic capabilities. Thirty years ago, most executives had secretaries and assistants. Today, they all better be well-versed in all the Office applications (Word, PowerPoint, and Excel) because many parts of their jobs require a more hands-on approach to reviewing day-to-day operations.

The attitude displayed by those keeping up with new developments like Social Media provide an example not only for their peers, but also for their subordinates. By using different tools and technology on a first-hand basis, executives can get a clear picture on how effective new technologies can be as well as knowing which ones to steer clear from.

There is a fallacy evident in many organizations which presumes once you are a manager you can manage anything. With technology, you cannot begin to manage until you become familiar with the capabilities as well as the limitations of resources within each level of technology at your disposal.

As Bob James, the former National Director of Telecommunications for Arthur Young, now Ernst & Young, once said about understanding the management of telecommunications, *"You either know it or you don't!"* *"There is no in between in managing this area."*

In organizations which have executives who are not skilled in the technologies they manage, the performance of the systems as well as the support personnel associated with the systems is mediocre, at best. The most dynamic organizations which utilize technologies to their fullest are those having executives who were not afraid to get their "hands dirty."

This philosophy implies that the CEO should have a solid technical background coupled with a business background which will provide a greater understanding on how to get more out of the technical resources available to attain organizational objectives.

It leads to the second cornerstone of the Strategic Triad of Technology Management:

YOU MUST UNDERSTAND THE TECHNOLOGY AND ITS ORGANIZATIONAL APPLICATIONS IN ORDER TO MANAGE IT

Knowing how to apply technology by itself is not the final skill the executive should possess. There also needs to be a skill focusing on an understanding of where the organization is headed.

Understanding the application of strategic planning is a must. The executive should be able to answer the following types of questions:

How can a communication based information system provide a competitive edge for the organization?

How can a communication-based information system or Social Media tool be used as a delivery system for new products or services offered by the organization to internal as well as external customers?

What types of technologies could be implemented to save costs or streamline manufacturing, order processing, or customer service?

Where is the organization going to be in five years? What systems and services need to be put in place now to support growth to that point and beyond?

In order to be able to ask and answer these broad questions, the top executives must study and directly contribute to the strategic business plan of the organization.

With the knowledge of the direction the organization is headed, executives can plot a more accurate course of action for the planning and implementation of technologies as a strategic advantage. This knowledge also insures that the systems and services which are being reviewed and implemented contribute to the attainment of the overall strategic objectives of the organization.

With this, we secure the final cornerstone of the strategic triad of technology management **(See CHART 9-5)**:

YOU MUST KNOW THE STRATEGIC DIRECTION OF YOUR ORGANIZATION IN ORDER TO APPLY THE RIGHT TECHNOLOGY

The skills of understanding and managing technological change, as well as understanding and being directly involved in establishing the strategic business direction of an organization, should be augmented by the skill of knowing how to lead people. Diverse, technology based resources must be managed, but personnel must be led in order for the organization's objectives to be attained.

CARLINI-ISM: *"Creative people are scarce and do not come cheap. When you pay peanuts you get monkeys."*

RAMMING SPEED!

Some leaders of enterprises just do not understand what new skills are needed in order to make a new, creative idea happen.

There are other organizational leaders that are reminiscent of captains of the old Spanish Galleons. They view the person in charge of directing the organization's implementation of technology and automation to its business needs as nothing more than a timekeeper beating the drum to get the oarsmen to row faster.

"We want to move ahead into new markets!" the captains will say. *"Pull the oars faster!"*

Technology does not work that way. Sometimes, you have to replace the technology in order to gain momentum on the market. Sometimes, you have to replace the people in charge because they do not understand how to apply the new technology. In recent years, this has happened in many organizations.

Some recent social media tools have not been as robust as what some thought. It is critical to distinguish between the tools and the toys.

CARLINI-ISM: *"Taking something which has been proven in one industry and applying it to another industry can accelerate the solutions to both simple and complex problems."*

CIO, THE ORGANIZATION'S TECHNOLOGY NAVIGATOR

As time goes on, you will see organizational ships sink, go off course, proceed on course at a snail's pace, and a few that proceed on course gaining tremendous speed in addition to gaining respect from their competitors. All of them will have had their captain, the CEO, and their navigator of technology, the Chief Information Officer (CIO) or some refer to the Chief Technology Officer (CTO). Both "officers" will have had their management methodologies when describing their successes as well as their executive and engineering excuses when discussing their failures.

With the proliferation of communications based information systems into all types of organizations, the need to assess the current adaptability as well as the future flexibility of organizational management to these new concerns is a major factor for those seeking a productive work environment.

CHART 9-5: STRATEGIC TRIAD OF TECHNOLOGY MANAGEMENT

YOU MUST KEEP ABREAST OF TECHNOLOGY
YOU MUST KNOW THE STRATEGIC DIRECTION OF YOUR ORGANIZATION IN ORDER TO APPLY THE RIGHT TECHNOLOGY
YOU MUST UNDERSTAND THE TECHNOLOGY AND ITS ORGANIZATIONAL APPLICATIONS IN ORDER TO MANAGE IT

CHAPTER REVIEW

QUESTIONS TO DISCUSS AND REVIEW

1.   What are the five levels of technology on the **TECHNOLOGY MAP**? Why is determining risk for each applied technology important to understand?

2.   How do you determine the mix of technologies to apply to an organization? Are they the same if you want to lead the industry or if you just want your organization to be part of the pack?

3.   Why is having a well trained workforce so important to all types of organizations?

4.   What happens when people are not rewarded for their work? If you are in charge of an organization what would you do to insure recognition?

5.   Why is it important to set an example for subordinates? What are some important issues?

CARLINI-ISMS

"You must understand the usefulness of the technology infrastructure servicing your enterprise."

"Applying network based information systems to sustain a competitive lead is not a one shot deal. It's a continuous process."

"Setting the standard is the sign of an industry leader. Playing catch up or succumbing to providing mediocrity is the sign of a member of the trailing pack."

"If an organization is stagnant, all it wants is someone to maintain the current systems – in effect, a network janitor."

"A well-trained workforce is a good hedge against the competition."

"Creative people are scarce and do not come cheap. When you pay peanuts, you get monkeys."

"Taking something which has been proven in one industry and applying it to another industry can accelerate the solutions to both simple and complex problems."

[10]

RFPs & PROCUREMENT

"We shape our buildings; Thereafter they shape us" – **WINSTON CHURCHILL**

W hen it comes to any executive position in real estate, technology, infrastructure or government, chances are you are going to have to review contracts and purchase both goods and services. Most people's decisions hinge on "**PRICE**" and many look to *"Who has the cheapest cost?"* as their main criteria. It was the wrong approach in the 20th century, and it definitely is the wrong approach today in the 21st century.

Price is not everything, especially when it comes to building and supporting next-generation high-tech real estate as well as regional infrastructure.

CARLINI-ISM: *"There is no such thing as a new $5,000 Rolls-Royce. You get what you pay for."*

CARLINI-ISM: *"Conversely, if you only have $5,000 to spend, there is no such thing as a Formula One Yugo."*

This chapter covers a proven twelve-step procurement review process for RFPs (Request-For-Proposals) to get away from "**PRICE**" as the overwhelming, sole procurement criteria for product and services selection. If you are looking at any standalone computer system, communication system, server hardware, software package or any other technology, this is the approach you should adopt and apply.

There are twelve criteria in the procurement review process for assessing technology before you buy it. Price should not be the single determining factor. Adding these other criteria makes for a better, more-informed decision that adds resiliency to the final product. By using this process, it forces you away from using "price" as the single most important factor in selecting equipment and services.

In evaluating choices for hardware devices and network infrastructures, there is a wide selection with premise based, central office based systems, and wireless systems as viable alternatives along with other devices, components, and network services used for cloud computing and edge technology. The best choice is selected when you are "systems literate" in the area of computer hardware and network communications.

CARLINI-ISM:
"You don't have to be an expert. You just have to ask a couple of detailed questions so the vendor thinks you are an expert and believes he/she better not try to sell you a bill of goods."

This RFP approach was initially developed for a client, AT&T, and taught for years at Northwestern University in my technology management courses in both the undergrad and Executive Masters programs as a way to get a broader and more objective perspective when buying technology and services. Learn it, because it is pragmatic. Use it, because it works.

The procurement approach for technology in infrastructure and how to review alternatives with a better metric using a new model is an important tool. **(See CHART 10-1)** Replacing old rules of thumb with new pragmatic questions that are divided into critical areas is emphasized. A realistic approach in evaluating current options is provided as a metric for consultants as well as customers to use.

CARLINI-ISM: *"In comparing options, you always want to have an objective yardstick."*

This sample Request For Proposal (RFP) Guidelines package gets back to the basics as well as establishes some new evaluative methodologies focusing on critical elements to provide a strong foundation for comparing different vendor offerings. It forces you to create a broader perspective on what you are buying and guarantees you will not evaluate and select solely on "price" as a single deciding criteria.

All you have to do is remember R, F, and P to do a good analysis of technology for procurement:

CHART 10-1: THE "RFP" PROCUREMENT CRITERIA APPROACH

AS EASY AS "RFP"	CRITERIA
R	REDUNDANCY
	RELIABILITY
	REDUCED OPERATING COSTS
F	FUNCTIONALITY
	FLEXIBILITY
	FAMILIARITY
P	PERFORMANCE
	PERSONNEL
	PRICING
ADDED CRITERIA	
R	RESOURCES
	RISKS
	REALITIES

Applying technology to any enterprise is a continual process. Having a more objective and comprehensive methodology to review and select equipment, services, and any cloud computing component can only create a better, cost-effective approach on upgrades and new capabilities.

REMEMBER THE THREE R'S WHEN EVALUATING ANY SYSTEM

From a traditional sense, there are several areas to evaluate any type of communications-based information technology system for both large and small organizations. It will be easy to remember these criteria because they all begin with R, F, or P. The first three areas are:

RELIABILITY
REDUNDANCY
REDUCED OPERATING COSTS

RELIABILITY

As more organizations rely on several forms of network infrastructure as a critical link to internal as well as external communications, reliability becomes a primary concern. There are mission critical networks and applications which have-to-have high reliability. They say one-out-of-three applications is mission critical. In a couple of years, the ratio will be one-out-of-every-two applications will be mission critical. These differentiating factors will be important to incorporate in your evaluation and selection process.

The levels of reliability for enterprise wide communications are only as good as the equipment they run through. An organization should evaluate its business needs first, prioritize them, and then determine what type of system and network services are required to support those needs as well as supporting day to day operations.

If you are managing a business campus, the same issues should be addressed by you and your staff.

If you have a mission critical application, you should also be looking at "five nines" worth of "uptime" reliability. (**See CHART 11-1** in Chapter 11)

From the customer's perspective:

What is needed to support current applications as well as future applications?

Is a critical downtime measured in hours, minutes, or seconds?

What is the cost of downtime to the user in a full or partial system failure?

How stable are the vendors that will be supporting the system?

What are the procedures for continuing business through a disaster?

From a hardware standpoint:

What is the mean time between failures (MTBF) for each major component of the system? (i.e. Processor, power supply, line cards, trunk cards)

What about a distinction between major and minor failures?

Define a major failure as well as a minor failure.

What is the aggregate MTBF for the whole system?

What about mean time to repair (MTTR) on each component?

From a vendor standpoint:

What type of support is included?

What level of services can they offer?

Is there a business day (9 5) service as well as a twenty four hour (7 day a week) response?

Are there guaranteed response time services that are offered?

Is there a way to measure that performance to insure that the customer is getting what they are paying for?

Is there a penalty provision if the performance criteria is not met?

CARLINI-ISM: *"No system or technology runs at 100% reliability. Even the phone company has down-time within its network."*

REDUNDANCY

No matter how reliable a system is being sold as, failures do occur. The management question becomes, what areas can the organization afford to back up? What can it afford not to back up? When a failure does occur, what back up capability is activated? What type of redundancy is built into the system and what levels of redundancy are available at an extra cost?

If you are in a mission critical environment, which today accounts for one-out-of-every-three applications, you better have a totally redundant system with complete back-up. Mission critical applications are growing in organizations to the new ratio of one-out-of-every-two applications within a couple of years.

This is a critical concept to understand as well as fund. It also makes you focus on changing these very traditional building approaches where you have one connection into the building from one central office. If you have a mission critical application, you need to have two separate connections going to two different central offices for communications redundancy. (If you don't – fix it!)

The same "redundancy" applies to power sources as well. You should be on two separate power grids if you are supporting mission critical applications.

(Think of this type of redundancy at home as well, if you are working as a telecommuter.)

At a processor (computer) level:

Do you have fully redundant processors?

If not, what is the cost for a second processor?

Can you create a mission-critical system (one that does not have a single point-of-failure)?

Is the second processor a hot standby or does it need to be manually booted up?

How much time does it take to switch over to the other processor and is it transparent to the user?

What types of calls are dropped in a switch over of processors?

Will a software failure in the first processor be mirrored in the second processor?

At the power supply level:

Does the system have dual power supplies?

Does the system provide battery back up?

For how long and at what incremental cost for one hour? Two hours? Three hours? Eight hours?

Is there a difference in the length of coverage if failure occurs during normal working hours compared to after hours, weekends, or holidays?

Is there a diesel generator to back up the system?

How long can the generator operate?

What special requirements do you need to install and operate the diesel generator?

Is natural gas an optional power back-up? Same questions for the natural gas back-up as the diesel generator.

What happens in a power failure from the serving utility?

Is there a second power feed from a different part of the power grid? Any alternative power sources?

As for phone systems, do all the phones lose power?

How many phones should be backed up (powered) from the central office?

What is the additional cost?

Here are some seldom-asked critical questions:

What is the cost to install redundant capabilities compared to the cost of one hour's worth of lost business calls?

What is the cost of one hour's worth of lost calls to the business?

What is the revenue made within an average hour on an average day and on a peak day?

Does that change depending on the season (what season is the "peak" season)?

For example, calls for orders placed in the holiday season are significantly more in value than the cost of calls for orders in the summer in several industries including the retail industry. Does your organization realize this? Has anyone figured out how much revenue is generated in one hour within the call center?

In another example, an airlines reservations center processes a number of revenue-generating calls each hour. How many calls do they receive in different seasons? What is their "peak" season? What is their "peak" hour?

At well-run call centers they have this type of information at their fingertips. They know what the value of each call coming in is worth in peak season like the holiday season, or at a low-volume season. They can calculate what an hour's worth of downtime costs them in any part of the year. Can you do this at your organization?

CARLINI-ISM: *"If you cannot determine what downtime costs are on a per-call or per hour basis, you are not managing your call center or any network effectively."*

At the circuit card level:

What type of redundancy is at the line and trunk card level?

How many lines (trunks) are on each card?

From a personnel standpoint:

What about the need for redundant skills that is necessary to manage the systems? Should anyone be cross-trained?

Do you have people on staff or do you rely on an outside third party for support?

REDUCED OPERATING COSTS

What are some of the cost savings if the organization upgrades to the new system? Will the new system be able to reduce staff or offload routine work so that the staff can be re positioned into more challenging jobs?

Will the costs for moves, adds and changes be affected? Positively or negatively?

Many times buyers will review the upfront costs, but not analyze the ongoing costs. ***(See Pricing section.)***

What about ongoing maintenance costs associated with the system?

Can maintenance agreements be broken down to cover pieces of the system? (i.e. Processor, peripherals, terminals)

What about the ongoing internal costs associated with the administration of the system?

In what areas will the new system increase costs? (i.e. Higher administrative costs for new systems administrators, technicians) can these costs be justified? How?

What about the cost for spare parts to keep on-hand? Components? Circuit cards, phones?

(A little known rule-of-thumb: telecommunications requires a spare inventory that could range from 1%-3% of the total organization's telecommunications budget.)

What about the costs associated with improvements within the facility to accommodate the system? For example, floor space needed (cost per square foot), air conditioning, battery back-up room, additional power requirements, and special construction costs.

What about the cost for an unexpected large growth due to adding a new division, consolidation of offices, or acquisition of another company? What about a downsize?

Do you support BYOD (Bring Your Own Device)? Are people using their own Smartphones within your organization? How are you insuring this is being done correctly?

In The New Millennium: There Are Another Three R's To Remember

Additional areas for evaluation criteria can be found in the following:

RESOURCES
RISKS
REALITIES

RESOURCES

From a strategic level, network and communications infrastructure should be viewed as a resource, not as an expense. A strategic resource to any organization in the 21st century and beyond will be its communications infrastructure. Once an organization acknowledges that its communications infrastructure is a strategic resource, the potential network and systems solutions cannot be evaluated using a commodities (*"BUY THE CHEAPEST"*) approach.

From a tactical standpoint, how many resources will be needed to manage, administer, and maintain the communications infrastructure?

How can the system be used to help compete against time?

How can the system be used to help get products to the market quicker?

How can the system be used to respond to customers quicker?

What kind of measurements are in place, if any, that determine the impact of the technology on sales, customer support, manufacturing processes, and other organizational areas?

RISKS

Any time that you install a new system, there is an element of risk. What can be done to minimize the risk and maximize the potential applications of the new system?

Is disruption to service due to installation or temporary failures important? How can this risk be minimized?

How many stations are you going to bring up? How complex will the installation be?

Are any other systems being installed and activated at the same time? (For example, voice mail, automated attendants, and automated response units)

Will you need outside help to install the system(s)?

What type of coordination will be needed?

What type of training will be required to manage the system and how many internal people should be trained to manage the system?

What is considered adequate support for the installation?

What about third-party offerings? What hosted services are available?

With many applications being "mission critical" today, you better insure the critical ones have adequate redundancy and back-up capabilities. Mission critical applications need to be clearly reviewed to minimize risks associated with their operation as well as their back-up capabilities to insure business continuity.

REALITIES

Many rules-of-thumb that people use become quickly out-of-date, especially in the area of network communications. The realities of evaluating systems are sometimes not considered.

The initial upfront cost is used to compare solutions for many organizations, but what about ongoing service and support for the life of the system?

What are other issues besides cost that are important to the user?

How can these be measured and translated into criteria for evaluating different system solutions?

Are there any growth limitations on the system? What is the scalability of the system? Where does it max out in size for ports for outside trunks and stations?

Is there enough processor power to handle peak operating times?

From a service standpoint:

How many systems are in place today within a twenty mile area? How many lines do they serve?

How many trained service technicians are available within that twenty mile radius at any given time (24 hours a day)?

How fast can problems be resolved if a problem occurs during normal working hours, after hours, weekends, or holidays?

Service availability may not be considered upfront and yet for the lifetime of the system, service and support is going to be key to the success of the system supporting the organization.

From a replacement standpoint:

What is the reality of getting a residual amount of money for a used system on the secondary market?

What is the real salvage value of the system after five years? Seven years? Ten years?

What are the associated costs to get rid of a used system? (That's right. An old system may be worth nothing in salvage value and it may actually cost you money to get rid of it.)

Some systems are worthless after five years and others have a salvage value because they are still in use within the market. Don't discount the value of a system that might be higher upfront, but will keep some residual value after five years.

What About The Three F's?

FLEXIBILITY
FUNCTIONALITY
FAMILIARITY

FLEXIBILITY

Any system that is acquired today must have the capabilities to be flexible to adapt to the dynamic environments of today's markets as well as being adaptable to tomorrow's challenges.

Can the system adapt to emerging technologies and services? At what cost?

From a hardware standpoint:

What has to be replaced or added to the system to keep it compatible to the latest technologies? At what cost?

Is the system proprietary or does it conform to some standards?

From a software standpoint:

What about software upgrades?

Are there any software upgrades provided during the life of the system?

What are the costs associated with software upgrades?

From a personnel standpoint:

How difficult is it to manage the system and is there a pool of outside people that are skilled as well as available if key administrators are lost?

FUNCTIONALITY

What are the existing and projected business applications and strategies of the organization?

Will the system be able to keep up with the organization's growth?

What about elasticity of the business - peaks and valleys of sales?

How can you apply the system to the organization's needs?

Are there any requirements that are better met by other means? What are the alternatives?

How do they compare in cost, risk, and benefits to the proposed system?

FAMILIARITY

All systems are "user-friendly." That's why user-friendly is not a real criteria. We discussed this at length in the course about buying and selling technology. I always had a standing offer to all the students to go to any conference and talk to the vendor about their system. I told the students the vendor would always say their system was "easy-to-learn" and "user-friendly." I told them I would give them a twenty-dollar bill, if they could find a vendor to tell them that their system *"was a bitch to learn."* I never had to pay off on that bet. That's when I pointed out to them that the criteria of "user-friendly" is not even worth putting on an RFP as no vendor is going to admit their systems are complex and hard to use. In reality, they are.

CARLINI-ISM: *"No vendor is going to tell you their system "is a bitch" to learn. They are all user-friendly."*

The user of any system must be comfortable with using all or most of the features of it. Whether a system has twenty features or two hundred features, it is only successful if the user can easily adapt to activating the capabilities of the system. With many users only using six to ten features, the total system functionality may be grossly underutilized.

What is the availability of training, information and documentation for the system?

What type of initial training is required? What is provided? Is there an initial cost?

Are there additional costs for more personnel to be trained?

What methods are used to train users that come onto the system at a later time after the initial cutover training has been completed?

Are user guides or system feature reference cards needed or available?

Do the users feel comfortable with the range of options for CPE (Customer Premise Equipment) equipment?

How can utilization of capabilities be increased?

Is the CPE equipment compatible with other telecommunication systems within the organization?

How long is the learning curve to understand how to use the equipment?

Is there a hot line for system users for learning the features during the initial weeks?

CARLINI-ISM: *"No system runs by itself. You need a systems manager or administrator."*

You need administrators as well as technicians to manage the day-to-day operations of any communications-based information system. If you are supporting BYOD, what are you doing to insure this is being run effectively?

WHAT ABOUT THE THREE P'S?

PERSONNEL
PERFORMANCE
PRICING

PERSONNEL

What may look like a cheap solution upfront, may grow into an ongoing budget concern as personnel costs needed to manage the system becomes evident.

> What are the management requirements for the system that you select?
> Does it need a full-time administrator?
> How many hours a week are needed by the systems administrator to manage the system, if not full-time?
> On average, how many technicians are needed to support the system at 50 station increments? (i.e. 1 at 50, 2 at 150, 3 at 300)

Can you get third party expertise to manage the system? At what cost? Are there options on third-party management (Business day versus 24/7 support)?

PERFORMANCE

What is the organization getting for the money that is being spent on this project?

Can performance be measured as it relates to supporting organizational areas of customer support, sales, and other areas?

> From a hardware standpoint:
>> What about blocking at the processor level? Although vendors claim that their switch is non-blocking, other components within a switch can block calls: tone generators, the amount of outside trunks, and other components.
>> What capabilities are built-in and what are add-ons?
> From a vendor standpoint:
>> Is there a built-in mechanism to measure the performance of the vendor?
>> Are vendors proactively managing the health of the technology so that they know there is a problem before you do?

PRICING

When reviewing any system costs, an evaluation should be done on upfront costs as well as ongoing costs. The costs associated with maintaining one system over another may be a significant factor that is overlooked in the initial review. Just comparing upfront purchase pricing is the sign of poor management.

Many organizations looking at implementing a new network infrastructure for Cloud Computing; a new building being built; or any other network capability, better look closer at their network designs.

Why? All the cabling needed to build the network should fit the lifespan of the building, not the lifespan of the technology hanging off of it. A lot of network architects, network designers and "certified" network engineers seem to be forgetting this fundamental design concept.

They also seem to forget the total costs involved in building a network, maintaining it, modifying it, and upgrading it.

LOOK AT THE TOTAL COST, NOT JUST THE UPFRONT COSTS

I discussed the *True Total Cost of Networks* (**TTC**) in several articles as well as a couple of presentations at national conferences.

Unless you factor all of these variables together when you compare solutions, it does not add up to the **True Total Cost (TTC)**.

(See CHART 10-2)

CHART 10-2: TRUE TOTAL COST APPROACH

TRUE TOTAL COST APPROACH FORMULA	
TTC = UC + (OC + OIC) + HC	
UC	*Upfront Costs*
OC	*Ongoing Costs (external)*
OIC	*Ongoing Internal Costs*
HC	*Hidden Costs*

Hidden Costs

Many times, costs that are not readily apparent are not added into the evaluation formula.

Consider the following "Hidden Cost" areas:
What about shipping costs for the switch and Customer Premise Equipment? A rule of thumb to use to determine approximately what the cost should be is:

What is the cost for a spare parts inventory? (Usually 1% to 3% of total annual telecommunications budget) If you spent $10,000,000 on a phone system, you may have anywhere from $100,000 to $300,000 worth of spare parts inventory that you must invest in and keep in storage.

What is the availability of spare parts?
How long of a wait is there for the worst case?

Where are spare parts located critical and non-critical? Customer site? Local service center? Regional service center? National service center?

It would be good to get a breakdown of equipment with these questions being answered. It is a lot different if a part has to come from a local dealer versus some regional or national warehouse.

How many levels of stock inventories are there before the manufacturer gets involved?

Many people must have missed the articles and presentations because few actually look at network projects using the **True Total Cost** approach.

By implementing their network and then having to go back and re-cable it later because it became prematurely obsolete, they have proven that there is another hidden cost involved in their shortsighted approach when cabling new and existing buildings: The Cost of Disruption.

THE COST OF DISRUPTION

There are many who will initially argue this "expanded" cost structure of TTC and say the customer won't pay for it. To that, I say, *"You need to sell the customer on why it is important to take all these costs into consideration and install the right network which is going to last longer."*

The design costs too much? You don't want to initially pull fiber along with the copper cable and pay for all the labor once, instead of twice? Why not? There is a legitimate cost savings there (just the labor cost alone is significant).

Not good enough? Then calculate the "Cost of Disruption." How much does it cost three to five years from now when the new building is all built out and occupied, and you have to upgrade the wiring throughout the building because you do not have the capacity anymore for all the applications needed to be serviced?

Move out the desks, break open the walls, pull cable from point A to point B, repeat 500 or 1000 times, disrupt whoever is working there and tell them it will only be a few weeks before everything will be upgraded. Put a price tag on that!

Calculate all the demolition and re-building. Calculate the additional drywall needed, the re-painting of the walls and other finish issues. Calculate all the lost productivity throughout the workspace.

What if it is a law firm or a medical office? How many hours are lost? It is not just the cost of materials and labor for the job; it is also the cost of disrupting all those professionals who are supposed to be working in the workspace.

Are you really saving money upfront by not pulling through that extra cable at the time of the initial cabling installation? I know you aren't.

Add to the Cost of Disruption the extra penalty of "Extra Cost to Upgrade" because the added labor cost. That could be enough for some accounting person to say no to the upgrade because it costs too much. The end result: A prematurely obsolete installation that has far less capacity than what it should have, making the building or venue less viable. **(See CHART 10-3)**

Look beyond the initial project stages. Look at what the longevity of the cabling infrastructure should be. Spending a little more money upfront can guarantee a longer life for the network infrastructure that you want to use. Don't forget the Cost of Disruption when you calculate the cost of the job.

CHART 10-3: TRUE TOTAL COST APPROACH (+CD+ECU)

TRUE TOTAL COST APPROACH FORMULA WITH ADDED COST OF DISRUPTION & ECU TTC = UC + (OC + OIC) + HC + CD + ECU	
UC	*Upfront Costs*
OC	*Ongoing Costs (external)*
OIC	*Ongoing Internal Costs*
HC	*Hidden Costs*
CD	*Cost of Disruption (if you go the "cheap route" and do not build enough for future growth)*
ECU	*Extra Cost to Upgrade (it's an extra – and unnecessary project. What does it cost?)*

CARLINI-ISM: *"Beware of hidden costs and the penalty costs like the "Cost of Disruption" and the "Extra Cost to Upgrade" when projects have to be prematurely rebuilt because you did not think long-term."*

What should you look for?

What type of financial analysis are you going to use?
Discounted Cash Flow (DCF)
Net present value (NPV)
Internal Rate of Return (IRR)
Discounted payback period (DPP)

What are the vendor specific, customer specific, and system support costs associated with all of the options?

(POWER, HVAC, FLOORSPACE for switch room, battery back-up room, parts inventory, SECURITY)

A discounted cash flow (DCF) should be run on the networking requirements as well as the hardware system(s).

ADDITIONAL QUESTIONS THAT WILL REVEAL A LOT ABOUT WHAT YOU ARE BUYING

Are you buying a system that can be expanded? Or, are you buying a barebones system that as soon as you need to add another five or 10 phones, it is a major upgrade? Or worse yet, you need a whole new phone system just to accommodate a couple of new users.

Here is a "must-have" question to be asked in the RFP when you are adding a phone system to your office:

What is the total cost to add on 5 phones? 10 phones? 20 phones? 30 phones? 40 phones? 50 phones? 100 phones? 200 phones? 300 phones?

What are the other equipment costs besides CPE equipment that are needed in the above examples? (Watch out for maxed-out systems where adding another five phones won't be the incremental cost of just the phones, you will also have to buy more supporting hardware like more circuit cards, an extra cabinet to house them and maybe even another power source to power up the cabinet. Instead of five phones costing you $500, you might find out you need to expand the system and that expansion costs $8,000 when it is all said and done.)

What about total installation costs?

Any special construction costs for equipment room/ battery room?
What about costs associated with wiring the system to the desktop?
Cable costs, connectors, terminations.

INTERNAL PERSONNEL COSTS

No system runs by itself. These are significant questions to ask and get answers from the vendor(s).

What are the cost for technicians to support moves, add-ons, and changes during day-to-day business operations? What about off hour installations - what is the premium paid?

What does it cost to manage the system? (Administrator to update passwords, levels of service, feature capabilities)

What are the loaded salaries for the personnel needed to run the system? For example, the system administrator's cost, technicians' cost, and anyone else needed.

COST OF FUNDS

For phone systems: What does it cost to purchase or lease the system versus the costs associated with a service agreement on central office based services? These are important questions to ask:
What impact does buying the system make on the organization's balance sheet?

What is the cost of capital and is that a factor to the organization? Some organizations are for-profit, others are non-profit. (That makes a difference when you do this analysis.)

CHAPTER REVIEW (See CHART 10-4)

The Three R's – Reliability, Redundancy and Reduced operating costs

The Three F's –Flexibility, Functionality, and Familiarity

The Three P's – Performance, Personnel, and Pricing

The Three R's for this century – Resources, Risks, and Realities of the global markets.

CHART 10-4: YARDSTICK FOR TECHNOLOGY PROCUREMENT

RELIABILITY	At what percentage of reliability is the service, 99.9%? 99.99? 99.995? How is this measured? What costs are associated with increasing reliability? (Cost/Benefit Analysis)
REDUN-DANCY	Can the system survive a major disaster or is there a single point-of-failure? Does the vendor have a system that can handle mission critical applications or is that better left off the cloud?
REDUCED OPERATING COSTS	Every system is going to proclaim they save operating expenses. Is there a significant difference between vendors' products?
FLEXIBILITY	How flexible is the basic platform to accommodate changes and/or additions to the initial applications?
FUNCTION-ALITY	What does the cloud provide as to capabilities? Are these functionalities easily adaptable to changes in the business?

FAMILIARITY	Every system is sold as "user-friendly" so that is not a good criteria to use. Familiarity is a better term to understand how the average user navigates into the cloud and is comfortable with utilizing the applications.
PERFOR-MANCE	Most systems are talking about performance in areas of scalability, elasticity, and other criteria, are there any unique performance capabilities tied to one vendor or vendors?
PRICING	Price is still a consideration in any purchase; it is only a part of the evaluation criteria instead of being 95% of the decision.
PERSONNEL	What people will the user still need to have on premise? A liaison person? A systems administrator? A SLA administrator? Any other resources?
RESOURCES	Are all resources off-site or are there requirements that might include on-site physical resources? What are the options?
RISKS	What are the risks in going to a third party, instead of keeping an application in-house? What about Intellectual Property or proprietary applications being put into the hands of a third-party?
REALITIES	What can be done with the cloud and what cannot be done? This sounds like a simple question yet many do not seek the answer to this and immediately look at the cloud as a "universal solution" to all their computing issues.

CARLINI-ISMS

"There is no such thing as a new $5,000 Rolls-Royce. You get what you pay for."

"Conversely, if you only have $5,000 to spend, there is no such thing as a Formula One Yugo."

"You don't have to be an expert.
You just have to ask a couple of detailed questions so the vendor thinks you ARE an expert and believes he/she better not try to sell you a bill of goods."

"In comparing options, you always want to have an objective yard-stick."

"No system or technology runs at 100% reliability. Even the phone company has down-time within its network."

"If you cannot determine what downtime costs are on a per-call or per hour basis, you are not managing your call center or any network effectively."

"No vendor is going to tell you their system "is a bitch" to learn. They are all user-friendly."

"No system runs by itself. You need a systems manager or administrator."

"Beware of hidden costs and the penalty costs like the "Cost of Disruption" and the "Extra Cost to Upgrade" when projects have to be prematurely rebuilt because you did not think long-term."

[11]

SELLING THE CONCEPTS OF INTELLIGENT AMENITIES

"Learning how to sell more than space today is critical for real estate professionals." – JAMES CARLINI

Selling in a new "connected" marketplace which cannot survive on a traditional approach to infrastructure and technology means breaking the traditional strategic approach to selling and leasing properties. The same advice goes for planning developments both from a public and private perspective.

Traditional real estate people understand how to sell "space" to potential tenants. They do not know how to sell intelligent amenities which add value and uniqueness to the "space." Some might argue real estate professionals have gotten more sophisticated, but I have seen too many go back to the *"cut-the-square-footage-price"* as a final competitive selling strategy when the going gets rough.

What are intelligent amenities? They are the technology-driven capabilities and value-added services which make a building unique in the market. When you understand that, you will not go back to the lowest common denominator of cutting "price per square foot" when dealing with buying or selling space.

Space is a commodity. Intelligent amenities, like broadband connectivity from diverse network feeds and carriers, are not. When they are offered as part of the real estate package, they can cut out over 90% of the competition. Stay away from selling commodities when you can sell unique, value-added services and amenities in real estate.

CARLINI-ISM: *"Be the best, or get passed up by the rest."*

New Concepts In Selling High-Tech Real Estate

Remember, we are past the industrial age. We are beyond the information age. Today, we are in the mobile internet age and that means people want communications delivered any time, any place and to anyone. If your building cannot support that, it doesn't matter if it has a marble fountain in the lobby.

Selling real estate has evolved because you are not just selling the elements of real estate anymore, you are selling the additional converged elements of technology and supporting infrastructure as well. Technology should be viewed as an investment, not as an expense within property management. From a selling standpoint, added capabilities which are part of an overall real estate package can narrow down the playing field substantially.

CARLINI-ISM: *"Technology should be viewed as an investment, not as an expense."*

In many organizations, the core business is intertwined with the platform of technology supporting it. Real estate firms are starting to observe this evolution within their own investments and those buildings and campuses which offer intelligent amenities are attracting a higher grade of tenant. In tighter markets, it may be the difference of keeping a tenant or seeing vacancy rates skyrocket up.

Cloud computing is a big application which many, if not all, organizations are looking into for their own core applications. The "reliability" chart found below **(See CHART 11-1)** gives you a clear idea to the levels of reliability of systems and services.

System and network availabilities are expressed with this type of chart focusing on uptime percentages.

Understanding this idea of "uptime reliability" and what it translates to as far as mission critical applications can give you added insights as to what potential tenants are looking for. (And what they are willing to pay for.)

Knowing how to discuss this area can be key when you are dealing with those who have mission critical applications which their organization depends on within their industry.

CARLINI-ISM: *"A good salesperson knows their product. A better salesperson knows their competitors' products as well."*

SELLING RESILIENCY

Resiliency is a big term many organizations are becoming aware of as they implement mission critical applications. Resiliency focuses on a speedy recovery from failures and disruptions. It is important because there has been a shift of disaster recovery concepts to business continuity in many organizations.

Disaster recovery means invoking an orderly shut-down of systems when a disaster strikes and when it is clear, the systems are turned back up and business resumes. A down-time is incurred by the organization and could be a couple of hours or a several days where business stops.

With a design focus of "Business Continuity", the attitude is when a disaster occurs, the business continues to function and nothing is shut down. "Disaster Recovery" and "Business Continuity" are very different strategies in the way they are designed and implemented for an organization. The costs to implement either one are different as well.

Working with a client like the Chicago Mercantile Exchange, they would want the highest resiliency for their networks. The reason is simple. They cannot afford any downtime with 100s of trillions of dollars of transactions running through their networks annually. What is an hour of downtime worth? You can calculate this value.

CHART 11-1: RELIABILITY/ DOWNTIME CHART

RELIABIL-ITY FAC-TOR	Uptime percent-age %	Down-time per year	Observation
"Two Nines"	99	3.65 days	This is WAY too much for any financial networks
	99.5	1.83 day	Better reliability, but still too much for most networks
"Three Nines"	99.9	8.76 hours	Depending on what you have to spend, can you live with this? Most cannot.
"Four Nines"	99.99	52.56 minutes	Less than an hour a year. Most could live with this.
"Five Nines"	99.999	5.26 minutes	Mission critical application? Here is where you need to be.
"Six Nines"	99.9999	31.5 seconds	Mission critical application? Even better.

When you can measure the cost of downtime, the cost for resiliency can be compared to what the cost of downtime is. If the cost of five minutes of downtime is $450,000,000,000, then spending $4,000,000,000 in money once to decrease it to an average of two minutes a year, is a bargain.

A supplemental checklist for resiliency for Service Level Agreements (SLAs) should include these criteria below. **(See CHART 11-2)**

The Four Levels Of Selling

Like anything else, there are low-skilled salespeople and highly trained salespeople depending on the product or service. You should realize where your expertise is being used.

The four levels of selling can be summarized by the table below. **(See CHART 11-3)**

CHART 11-2: YARDSTICK FOR AN SLA AGREEMENT

VENDOR PERFORMANCE ASSESSMENT	Vendor response performance, reports on various service response time metrics and overall customer satisfaction.
PERFORMANCE AUDIT TOOLS (1)	What are the ways the user can review performance and insure they are getting what they are paying for? (Standard response may be 4 hours, a premium response may be 2 hours. How is this being measured and stored for future review?) You must measure to insure you are getting your money's worth.
PERFORMANCE AUDIT TOOLS (2)	What about measuring surges, availability, and their impact on user throughput? Where are the potential weaknesses as well as the apparent ones? (System/service performance)
TRANSACTION LOGGING	Does the system track every transaction a user makes? To what detail? What types of reports are available for everyday and weekly reviews? What about special regulatory compliance reporting? (HIPAA, Sarbanes-Oxley) Are customizable reports available?
TIME-STAMPING	Is time-stamping of each user session, each transaction, and each resource used, available? Again, very important for trouble tracking and measuring resolution times.
TROUBLE REPORT MONITORING	Any troubles with system or services. How are they tracked and resolved? What type of reports are available? What reports can be customized?
USER-ID	Each user has a unique ID for audit purposes and how can they be monitored either on a real-time or a historical basis? Is user security 100%?

Source: James Carlini

CHART 11-3: FOUR LEVELS OF SELLING SKILLS

LAYER	LEVEL	DESCRIPTION
1	COMMODITY	Lowest level of sales, retail/ store environment.
2	SERVICES/EQUIPMENT	More competent salespeople. Formal training on products/ services.
3	APPLICATIONS/ PROJECT MANAGEMENT INTEGRATION	More specialized and technical salespeople. Focus is on broader integration of products vs. single product expertise.
4	BUSINESS PROCESSES	Complex and mission critical applications needing a strategic perspective.

1) **COMMODITY SALES** – Are the lowest-level sales and require the least amount of selling skills. It also applies to many outlets, the product or service is easily available and the customer perception is, *"I can get it anywhere. Who has the cheapest price?"*

The people at this level are typically minimum-wage workers and are of low competence. They may get paid an hourly wage and nothing else. Some of the jobs at this level would be entry-level.

2) **SERVICES/ EQUIPMENT** – Is the next sales level and it can pertain to a small scale vendors to large scale vendors with

many outlets. It may carry several product lines. The customer perception is, *"I can find some competitive products."*

The people at this level are paid more because they need to understand more.

They have more competence and probably have some formal training on the products. They could also be incentivized by a commission structure.

3) **APPLICATIONS/ PROJECT MANAGEMENT INTEGRATION** – Is the third level of sales expertise and it covers more sophisticated products and services. There could be several outlets, more specializations, and the overhead is greater. The customer perception is, *"I can find someone to integrate all the competitive products together."*

The people at this level are more competent, moderately more expensive, and probably are incentivized for their work and ability to sell work.

4) **BUSINESS PROCESSES** – Area has both few outlets and experts at a high level, scarce, and well-experienced. The customer perception is, *"I have to be in trouble to use this level or I need some complex products and services badly usually within some mission critical or strategic setting."*

There are few people at this level to choose from and they are highly-competent and expensive. They could be a strategic visionary, a specialist, or a generalist.

If you are in this market, there should be questions concerning the competition. From the buyer's perspective:

Are you buying at the right level?

From the salesperson's perspective:

Are you selling at the right level?

Where do you and your product/service fit within this structure?

How do you sell against those within your level and those within another level?

Can you sell against a different level successfully?

SELLING TIPS

Some selling tips I used to discuss in a marketing and legal class for technology projects can be found in the next chart. **(See CHART 11-4)** The good thing about these selling tips is they can also be applied to the real estate side as well.

CHART 11-4: SELLING TIPS FOR TECHNOLOGY

1	KNOW YOUR PRODUCT OR SERVICE – COLD.
2	KNOW YOUR MARKET - WHAT DO THEY WANT, WHAT DO THEY NEED?
3	KNOW YOUR COMPETITION - WHAT ARE THEIR WEAKNESSES, THEIR STRENGTHS AND RESOURCES?
4	KNOW YOUR COMPANY'S LIMITATIONS - YOU CANNOT BE EVERYTHING TO EVERYBODY - NOBODY CAN.
5	ALWAYS STUDY THE TOTAL PICTURE - COMPETITION, MARKETPLACE, AND OTHER OUTSIDE

	FACTORS TO KEEP IN TOUCH WITH REALITY.
6	DEAL FAIRLY - FORGET THE HARD SELL OR THE QUICK FIX.
7	STAY AWAY FROM PRICE AS A SINGLE BUYING CRITERIA. (AS A SELLER OR A BUYER). SALES ARE EMOTIONAL DECISIONS - NOT ALWAYS FINANCIAL ONES.
8	YOU MIGHT ONLY HAVE ONE CHANCE TO SELL THE CUSTOMER - DO IT RIGHT THE FIRST TIME.
9	KNOW WHEN TO WALK AWAY FROM BAD BUSINESS - GAIN INTEGRITY.
10	SUPPORT WHAT YOU SELL - YOU MIGHT UNCOVER MORE OPPORTU-NITIES AND CREATE ADD-ON SALES.
11	BE CREATIVE, FLEXIBLE, AND ADAPT-ABLE TO NEW OPPORTUNITIES.
12	CONTINUE TO LEARN NEW THINGS AND NEW CONCEPTS. IF YOU STOP - YOU MIGHT AS WELL RETIRE. THE MARKET IS TOO DYNAMIC FOR ROU-TINE WORKERS AND THOSE THAT CLING TO YESTERDAY'S SOLUTIONS.
13	BIG DOES NOT EQUATE TO GOOD. SMALL FIRMS ARE MORE ATTUNED TO SPECIAL MARKETS. DON'T DISCOUNT THE SMALL FIRM AND DON'T HOLD THE LARGE FIRM AS NECESSARILY THE BEST.

The Northwestern course was team-taught where I presented the marketing side and Russ Hartigan, an attorney (who is now a Circuit Court Judge), presented the legal aspects of information technology projects. This was an excellent way to teach a course focused on marketing issues and contract issues pertaining to the buying and selling of technology.

In classes, I always used to ask, *"Are you a salesperson?"* Many people would almost look in disgust and say, *"No way."* Then I would say, *"You are wrong. You are ALL salespeople because every time you go to your boss and request funding and resources you better know how to sell yourself and your concepts, if you want to get anything."* This started to hit home. People realized that no matter what, you were always trying to pitch ideas and concepts within their jobs and their organizations.

This was a great motivational spark in the classroom to wake people up to the fact "selling" is not a negative skill set. In fact, just the opposite is true. How many times have you tried to convince your boss a new endeavor would be good for the organization and were shot down? Understanding how to present things concisely and write well is a good executive skill set.

CARLINI-ISM: *"Executive skills include knowing how to speak well and knowing how to write well."*

Remember selling strategies with slogans like:

"SELL THE CHUMP THE LUMP"

"WE SELL VAPORWARE (SMOKE)"

"IT WILL ONLY COST YOU PENNIES A DAY" *and*

"OUR PRODUCT IS USER-FRIENDLY"

are all approaches and pitches which the market place has been exposed to, and have rejected. You cannot be a snake oil salesman and expect to survive in today's complex market.

If you are not motivated, your flimsy sales pitch will be easily detected by those who have negotiated agreements before. Customers are not naïve.

Working with Kenn J. Jankowski back at American Hospital Supply Corporation in the 1980s, he said something that stuck with me throughout my career. He set up many meetings with technology vendors and had me sit in. He used to say, *"Let's put these vendors in the frying pan and see what they really know."* He used to ask several very pointed questions about whatever they were selling and if they could not answer, he knew they didn't know too much about their product or how it would fit into his organization.

I remember one salesperson who "had to get back to us" with his technical support person because he did not know what SDLC stood for. At the time, IBM's SDLC (Synchronous Data Link Control protocol) was pretty basic stuff and you should have known that term if you were working with any large, corporate network. If you needed to bring in your technical person for that, how much did you really know as a salesperson selling equipment which used that protocol?

The story doesn't end there though. He did bring in his technical "engineer" on the next visit. The technical engineer told us he would have to get back to us because he had to talk to his regional technical support. They never were invited back. The reason? If they had to go to regional support just to answer such a basic question, how would they be in responding to immediate support issues if they actually installed their equipment into the enterprise? That is an important lesson even in today's technology marketplace.

CARLINI-ISM: *"When the vendor's technical knowledge and support is shallow, don't bet your company – or your job – on them."*

With those interviews, we summed up the vendor's ability to deliver what was needed. If they fell short, the right thing to do was to walk away from their offer.

This is another important piece of advice to walk away with: *You MUST believe in what you are selling.* If you don't, why should I? If you don't, why should your customers?

CARLINI-ISM: *"You have to believe in what you are selling. If you don't believe in it, why should the customer?*

Sales expertise is not learned in a 4-hour seminar or by reading a book. It takes time and practice like anything else to become proficient and confident. Learn from your mistakes as well as from the person before and after you. Try to target your market. Learn about the customers you are dealing with. Some deals are not good and should not be pursued.

CARLINI-ISM:

"Know when to walk away from bad business."

Sometimes, walking away from an "opportunity" is the best thing you can do. I remember several times when my decision to walk away from a big sales opportunity was the right one.

The first time was when I was a Director at Arthur Young (now, Ernst & Young). It was a big project to reconfigure the entire state of Wisconsin's networks into a more integrated network. Several firms were bidding on the project including AT&T, Boeing Computer Services, Arthur Young, and others. As the preliminary discussions progressed, the project director for Wisconsin started outlining what was expected in order to bid on the project. The first thing was to do a preliminary planning for the new network and submit it, and then they would decide who would move to the next step. I researched how much effort we were going to put in and saw quickly that we were going to have to do some initial configurations of their current network. We needed sophisticated software programs which could take all the traffic generated by the current network and manipulate all that data into some network configuration projections.

I went to my boss, the managing partner of the whole office and said, in order to compete for this job, we need to purchase some really sophisticated software that configures enterprise networks. When I told him what the software cost (at the time, $80,000) he thought that it was an expensive tool and we were not going to buy it.

That decision made my decision a lot easier. I decided we were going to fall out of the competition and not waste our time.

It was a big project, but it also cost a lot just to try to compete. I followed the rest of the project from a distance and asked one of the competitors how it was going. He said they still have not made a decision, but the amount of firms interested dropped down to two firms. His firm (Boeing) and AT&T.

He said he was so disillusioned from the project that he almost wanted to walk away. He told me they had already spent about $1,000,000 on the preliminary assessment and design and the customer still had not made a decision. They were already "on the hook" for $1,000,000 in consulting time, expenses and had no "locked-in contract" to show for all those tasks which were already finished. The customer was playing the eager vendors to do more and more for free.

At that point, I thought I made the right decision to walk away six months ago and saved our firm from spending over $1,000,000 that would have yet to be proved as money well spent. Instead, I spent time pursuing other client opportunities which made more sense to compete for the contract.

You must have a good idea of what you can deliver and for what price range before you take on any large projects. You cannot afford to waste resources. You also cannot be taken advantage of by customers, who want everything for free.

No matter what industry you are in, these are some words of advice to really adhere to: *Know when to walk away from bad business.* There are so many people I have come across in different projects and different professions throughout the years who sometimes forget this or because they are so focused in wanting to close a deal, they ignore this advice and pay the consequences later.

As soon as you mention this advice to senior marketing people or Top Gun sales professionals, they all groan and say, *"Yes, you are absolutely right."* It's like they have all seen this sometime in their past with colleagues or themselves and it has impacted them.

If you have less experience, learn this quickly and you will avoid some big problems. Your focus will mature faster than those who wait to learn the hard way.

There is an old saying, *"Learn from other people's mistakes."* This is very good advice when it comes to sales and marketing of technology.

People are still making mistakes as they are selling the latest technology into stadiums and other venues to support Smartphones and Tablets. It may be because they are not getting a good amount of training before they go out and sell to customers. We need to somehow remedy this approach because if the salesperson does a bad job, the solution may not work. If the solution fails, the technology is blamed as something *"that just doesn't work."*

We still have many people walking around claiming to be experts. Real experts in technology will perform a lot of analysis upfront to see how everything works together and develop a comprehensive solution fitting the problem. They will also tell you when they do not have a clear solution and need to invest more time in studying and prioritizing the complexity of the issues.

The pseudo-experts will walk in and with only a shallow understanding of the issues, proclaim they "have the solution." Beware of those who already know the solution, before taking any time to evaluate the situation.

SELLING IN TODAY'S REAL ESTATE MARKET

Trying to lease space in today's real estate market without these new intelligent amenities is like trying to sell a car without power steering, power windows, or a radio. Options which were rare or even non-existent are now considered to be required features by those who are looking for premium space.

CARLINI-ISM: *"Real estate firms are in for a big surprise. The buildings which lack solid intelligent amenities will lose first-class tenants and slip into second-class properties. It's happening already."*

As I observed several years ago, *"There are a lot of big egos in real estate, but if they don't follow this definite shift in the markets, they might as well concentrate on warehouses where the ultimate measure is still empty space."*

LEGAL PERSPECTIVE

In the legal and marketing course for information technology projects, the legal side was taught by attorney, Russ Hartigan. Russ was a corporation counsel for several municipalities in Illinois as well as an attorney in the area of insurance and municipal government issues. He added a lot to the course with pragmatic insights as to contract law, negotiating maintenance agreements, and other business law concepts as they pertained to applying technology to organizations.

One of the things he always pointed out was *"You always roll the dice when you take someone to court."* There was nothing like an iron-clad case, once it was presented in court. (He is now a Cook County, Illinois Circuit Court Judge.)

His practical insights from a trial attorney's perspective gave the students a much broader perspective than if they were taking the class only from a technologist. This team-taught approach is exactly what is needed today in presenting some of these multi-disciplinary concepts we have been discussing within this book. This approach works and works well. The material gets better coverage when two different perspectives from different professional backgrounds discuss the concepts in more of a panel discussion approach, than a single lecturer.

CARLINI-ISM: *"Real estate, infrastructure, and technology concepts are converging. They should be presented in a comprehensive manner in order to instill a broader perspective within students as well as professionals."*

CHAPTER REVIEW

CARLINI-ISMS

"Be the best, or get passed up by the rest."

"Technology should be viewed as an investment, not as an expense."

"A good salesperson knows their product. A better salesperson knows their competitors' products as well."

"Executive skills include knowing how to speak well and knowing how to write well."

"When the vendor's technical knowledge and support is shallow, don't bet your company – or your job – on them."

"You have to believe in what you are selling. If you don't believe in it, why should the customer?"

"Know when to walk away from bad business."

"Real estate firms are in for a big surprise. The buildings that lack solid intelligent amenities will lose first-class tenants and slip into second-class properties. It's happening already."

"Real estate, infrastructure, and technology concepts are converging. They should be presented in a comprehensive manner in order to instill a broader perspective within students as well as professionals."

[12]

CEO PERSPECTIVE: STRATEGIC VISIONS

"He who has the wheel, sets the direction." – **MALCOLM FORBES**

How should CEOs of organizations look at next-generation real estate for their organizations from a strategic perspective? What external platforms do they need for their organizations to maximize their facilities to increase profitability and sustainability in a global market? What new metrics do they need in order to assess, evaluate and understand alternatives and their value to the organization's core business?

These are just some of the real estate and facilities questions facing leaders of various organizations. People at the top need to understand this transformation and convergence which we are going through.

We are trying to transform government and corporate management to undertake the new task of setting a strategic direction as real estate, technology, and infrastructure converge to impact regional economic development.

There needs to be a whole new set of metrics developed as well as obsolete rules-of-thumb needing to be discarded.

In this chapter, I will highlight some leaders and their philosophies who I believe are good influences in developing a total executive approach to the multi-disciplinary challenges we face in this century. It is not good enough to be focused on a single discipline or managerial style. You have an MBA? An engineering degree? A real estate degree? So what?

Today's challenges require a much broader-based executive who can look at problems and react with a much broader set of skills than the typical MBA or real estate degree of the past. We do not want single-skilled people as much as we want multi-disciplinary people in leadership positions. **(See CHART 12-1)** The blue diagram represents four individual people versus the gold diagram representing one person with multi-disciplinary skills.

Again, higher education has not really changed its formula for management education in 30 years and yet the workplace, technology, and global market place have changed substantially in 30 years.

In a course I taught on Total Quality Management (TQM) and Six Sigma, we covered several methodologies of quality methods and when you look across all of them, whether it be Deming's 14 Points of Quality, or TQM or TCI (Total Continuous Improvement – or Kaizen), or even Six Sigma, they all start out with the same basic step – Assess the Current Environment.

CHART 12-1: TRADITIONAL VS. NEW MULTI-DISCIPLINARY SKILL

TRADITIONAL SKILL (SINGLE SKILL, 4 PERSONS)

INFRASTRUCTURE	REAL ESTATE	TECHNOL-OGY	BUILDINGS/ FACILITIES

MULTI-DISCIPLINARY SKILLS (ONE PERSON)

TECHNOLOGY
BUILDINGS/ FACILITIES
INFRASTRUCTURE
REAL ESTATE

As W. Edwards Deming once said, *"The problem is at the top, management is the problem."* This is true. Executive management sets the tone for the rest of the organization's culture. Executive management sets the example for the rest of the organization to emulate.

Did you know where W. Edwards Deming got his start? He was a quality analyst at Bell Telephone Laboratories back in the 1920s. Both he and Joseph Juran, another guru on Quality Management in manufacturing who initially worked at the Western Electric Hawthorne Works in Cicero, Illinois, worked on various studies and different experiments on improving productivity. Both got their start in the Bell System.

Deming and Juran found their big "claim to fame" when they went over to japan after World War II and helped rebuild Japanese manufacturing by instituting a lot of their quality methodologies. They became "gods of quality" when they came back to the United States and many people thought quality management initiatives were a Japanese management style when in reality, much of it was actually developed here.

Read some of their books on quality in organizations and it will add a great deal to your executive skill sets. Here is a summary of Deming's philosophy on Quality and how it impacts work and products. **(See CHART 12-2)**

CHART 12-2: DEMING QUALITY FOCUS & OUTCOME

FOCUS ON	OUTCOME
QUALITY	Quality tends to increase and costs decline over time.
COSTS	Costs tend to rise and quality decreases over time.

One of Deming's quotes really summarizes the reason for my book:
"To successfully respond to the myriad of changes that shake the world, transformation into a new style of management is required. The route to take is what I call profound knowledge—knowledge for leadership of transformation."

You need to know where you are at, before you can measure your improvement over time. Understanding the current conditions is the first step in making improvements.

Try to absorb some of Deming's observations as they are still relevant:

- *If you can't describe what you are doing as a process, you don't know what you are doing.*
- *Quality is everyone's responsibility.*
- *Any manager can do well in an expanding market.*
- *Lack of knowledge......that is the problem.*

A good leader has to have a set of skills which address both strategic and tactical levels of running an organization. He or she has to understand the tools required to make fair assessments when reviewing at new building sites or existing space for their staffs.

Many books have been written about different approaches and philosophies of various chief executives. Many executives have given different managerial quotes trying to encapsulate their philosophies and their beliefs. This book is not about corporate management or leadership, but having a chapter address leadership and setting strategic direction as it relates to assessing and deciding on what mix of technologies are to be applied to the organization is critical.

MANAGEMENT VERSUS LEADERSHIP

Too many times, management and leadership are used to define the same thing. There is a difference. Management is not the same skill as leadership and there needs to be a better delineation between the two:

CARLINI-ISM: *"You manage resources, but you lead people."*

You do not "manage" people. People do not want to be managed. When it comes to graduate "managerial education", I think most MBA programs fail to include some leadership courses. We need to re-think the curriculum and add at least one course to remedy this.

The necessary skill of leadership is one which can be seen as very important in today's complex global business environment. Too many schools teach the idea of being a "team player." Teams and team dynamics are good to understand, but you don't get a lot of strategic initiatives done by committee.

CARLINI-ISM: *"A leader can be a manager, but a manager may not be a leader."*

What needs to be taught is how to be a "team leader." By admission at almost every job interview, we have enough *"Team players"* applying for jobs. The standard question always asked is, "Are you a team player?" And the applicant's standard reply is always, *"Yes, I'm a team player."* What we need today, as well as tomorrow, are "team leaders."

In classes on technology management, I used to ask people about going on interviews and being asked the question, *"Are you a team player?"* The standard answer is, *"Yes, I am."*

I used to tell people to answer: *"No. I'm not."* And then after the interviewer would come back up off the floor from shock, to add, *"............I am a team leader."*

"Team players" aren't all we need. We need good team leaders and few grad schools were thinking about that.

If they were, do you really think we would have had all these corporate problems in the last two to three decades?

Many qualities were distilled down into clichés when people were taught interviewing techniques for seeking management positions. Seeing repetitive things on resumes like, *"I am a self-starter"* became meaningless as to someone's skills and motivations. I did see one I thought really bucked that trend and it came across with more impact. The applicant said he was a "self-finisher." Everyone can start a project, but how many can finish it without having to get help from others? It was another example I gave to classes to think about when going out for interviews.

So you're a self-starter? Big deal. That should be a given for any MBA or executive management job. Being a "self-finisher" – that seems to be spotlighting a better skill or quality.

We need to instill leadership qualities into people if we want them to lead organizations, teams, and make strategic decisions on development endeavors. We also need to recognize them and reward them when they perform well. No one stays at a place that does not recognize their work efforts.

CARLINI-ISM   *"If an organization fails to recognize talent, it starts to lose it to the competition."*

SETTING DIRECTION THROUGH LEADERSHIP

Malcolm Forbes once said, *"He who has the wheel, sets the direction."* For whatever reason, this quote has always stuck with me and in my thought process. It really resonates. It speaks volumes. It provides a good visualization.

Because he was an avid yachtsman, Malcolm Forbes had several phrases and words of wisdom which paralleled handling a ship. Another appropriate Forbes' observation for these complex economic times and lackluster CEOs is, *"Any fool can handle the helm in calm seas."*

I think Malcolm was right. There are some CEOs who coasted along for a long time, but as soon as a storm whipped up, they failed miserably. Their projected image of leadership washed away. They should have never been promoted into those "head of corporate navigation" positions in the first place.

Another leadership quote stands out in my mind and I have used it for a long time. It was first said by ITT Chairman, Hal Geneen. It is a good summarization of managerial effort and leadership:

"In business, words are words,
Explanations are explanations,
Promises are promises, but
Only performance is reality."

Wow. These are good words to live by. They are rich in wisdom and insight.

There are a lot of "big talkers" in all industries, including real estate. If you are a corporate leader, you need to be able to back up what you say and take action.

CARLINI-ISM: *"Only performance is reality. Everything else is just glossy rhetoric and slick slogans."*

EMPOWERING THOSE AROUND YOU

Let's debunk another "pop business management" concept. Empowerment doesn't work. You have to give authority when assigning responsibilities. Once you give someone else the responsibilities and authority to do something, you also have to recognize them when they do a good job.

CARLINI-ISM: *"Empowerment? Spread the responsibilities, but then spread the recognition and the rewards as well."*

The CEO (Chief Executive Officer) as well as other key executives must be able to coordinate and navigate people, technologies, systems, and other services in a direction providing the organizational ship the shortest route to attaining its strategic objectives, within the constraints of the organization's resources. They must be creative and know how to successfully apply technology to business needs. It is a complex task and is a difficult one to manage.

The CEO must also set an example by making an effort to constantly enhance their skills and knowledge. These traits should also be picked up by the other key executives and the local political leaders. The Strategic Triad of Executive Skills should be adopted. **(See CHART 12-3)**

CHART 12-3: EXECUTIVE SKILL SETS FOR ADOPTING AND ADAPTING TECHNOLOGY TO THE ORGANIZATION

SKILL SETS	REQUIREMENTS
BUSINESS	(CEO) You must know the strategic direction of your organization in order to apply the right technology (or technologies). (Property owner) You must know the strategic directions of your customer base in order to understand what to offer in your properties.
TECHNOLOGY	(Both) You must keep abreast of technology to understand what to select and apply. (both internally and externally) Don't just rely on a subordinate or vendor to do this.
MANAGEMENT	(Both) You must understand technologies and their organizational applications in order to strategically apply and manage them over time.

Choosing the wrong vision or head navigator will lead to moving the organization in technological circles. No real progress will be made.

It can get worse. The wrong decisions can move the organization in the opposite direction from where they should be headed with all of the applied telecommunications and information technologies acting as a strong tailwind accelerating the organizational ship into disaster instead of success.

In vendor organizations, the needs are similar. You must have someone who can be a visionary and sell that vision to the customer base. He or she must also understand the trends of the market, be able to set the pace of the organization and direction for product development and application marketing.

The necessary skill of leadership is one which can be seen as very important in today's complex broadband and digital systems environment. High turnover through mergers, reorganizations, poor management, and better opportunities has created a need for far better leadership skills which can positively address the demands of today's knowledge worker as well as shape a positive commitment to the organization's future.

CEOs must set an example by making an effort to constantly enhance their skills and knowledge because they are the "Head of Navigation" for their corporate ship moving in the trade routes of global business. They are the role models for their subordinates.

The problem is, few executives have made the connection of applying real quality approaches with applying technology to the enterprise. If they did, they would have a continual ongoing program to review AND implement new technologies into the fabric of the infrastructure that supports the organization.

If this happened, we would not have seen the dramatic decrease in technology sales in the early 2000s. Contrary to some of the pseudo-experts, IT budgets should have never been slashed.

CARLINI-ISM - *"You don't get Superman by offering Jimmy Olson wages. Serious people get serious money."*

If anything in times like that, those leaders who are still spending are creating a huge gap between themselves and their "wait-and-see" competition.

Many executives and their staffs are afraid to take any real steps in applying new technology. They have gotten complacent in their implementation and management of cutting-edge technologies.

Some executive search firms have exasperated this situation by placing and replacing these "corporate cattle" at job-after-job instead of looking at fresh faces and discarding the ineffective "must already be a CEO" criteria that most of them use.

In technology applications, what used to have a five- to seven-year window is now three-to-five years. Three- to five-year windows, now have a lifespan that is now more at twelve- to twenty-four months. A standard practice of executives taking a precautious "wait-and-see" attitude is now foolhardy. The strategic decision game has accelerated and new skill sets as well as an aggressive, decisive manner are mandatory, not a "hoped for" in key executives faced with entering windows of opportunities which are now opening and closing at a much faster rate.

APPLYING THE THEORY OF REALITY AS WELL AS SOME OTHERS

Using Robert Ringer's Theory of Reality from his best-selling book, **WINNING THROUGH INTIMIDATION,** we can set our pragmatic managerial perspective on the right course:

The theory emphasizes, first of all, that reality isn't the way you wish things to be, nor the way they appear to be, but the way they actually are. Secondly, the theory states that you either acknowledge reality and use it to your benefit, or it will automatically work against you.

If you use this as a cornerstone in your everyday management practices, you will remain on course. Ringer's book was one of the first I read right after college and I felt his book was better than most of the MBA courses I sat through. It is still relevant.

CARLINI-ISM: *"It's time to be politically accurate and NOT politically correct."*

Dealing with reality and being politically accurate can only make your decisions more accurate and focused on the right solution. When you start to cloud up your perspective and perception with flowery euphemisms, corporate-speak, and other obtuse assessments, that's when you fail to address the problem(s) at hand. Instead of decisive leadership which moves the organization forward, you slide down to "feel-good" management where everyone is complacent, no one rocks the boat, and nothing moves forward. Those organizations get overrun, bought out, or just die on the vine.

Another theory that you may want to remember is my *Impatient Theory.*

If people try to tell you that you are being impatient when it comes to getting a promotion, raise, or change of projects, chances are you are getting shuffled around. Waiting around does not get you anywhere. Let someone else be patient. Take action.

Keep a sense-of-urgency on what you do. Attack a project from several different fronts. If one direction becomes a dead-end, pursue the project from a different angle. Always keep moving! Think of yourself as General George S. Patton – *ATTACK! ATTACK! ATTACK!*

CARLINI-ISM: *"Take action. Let someone else be patient. Waiting around does not get you anywhere."*

Something will happen. Remember, everyone has a limited number of DARTs in business, as well as in life. Use yours wisely and aim well. Get things moving! Be a decision-maker and not someone who wastes a lot of time trying to make a decision.

If you have time, pick up some books like, **Patton on Leadership** (Axelrod), to read about leadership concepts. I used this book as another text in some of the management courses I taught. I am sure not all professors would agree with me in using Patton as an example, but I thought my students should get a good taste of pragmatic leadership which was successful in real life and not give them some untried, theoretical concepts.

General Patton had a pragmatic focus to get the job done. He had to – he had to win a war. He did not go for excuses or second-guessing. He was a master at logistics and moving huge amounts of men and materials under adverse conditions. One of his many quotes, *"Lead, Follow, or get out of the way"*, hits home pretty fast.

You could really improve your leadership qualities just by reading about Patton's actions, quotes and letting them sink in. Of course he is different than a Malcom Forbes, W. Edwards Deming, or Hal Geneen, but his leadership perspective is one you should adopt or at least blend in with your other leadership skills:

> *If everyone is thinking alike, then somebody isn't thinking.*
> *Always do everything you ask of those you command.*
> *Never tell people how to do things. Tell them what to do and they will surprise you with their ingenuity.*
> *Do your damnedest in an ostentatious manner all the time.*

His quotes could go on, but as suggested, do some reading on your own on him and the others. It will add a lot to your leadership skills. We could use more people with a Patton leadership mentality in today's business world as well as in government. (Yes, I was in the military and I believe it did me some good.) You get leadership training in the military that you won't get in any MBA curriculum.

Some "managers" like to second-guess their leadership. That is a bad idea. The same goes for boards of directors. You might be the president (or Chairman) of the board. The organization is depending on you to make a decision. A quick decision and not one that takes months and months to analyze. Many times, you must make a decision without having all the facts in front of you. If it is a crisis, do you really have time to gather all the facts before taking action? The honest answer is NO. You must learn to live with your decisions and no one can second-guess you or take a Monday morning quarterback review of your real-time decision.

So how much information should you have before you make a decision? 20% of the facts? 40% of the facts? Over 50% of the facts before you make a decision? The more time you take to gather the facts, the less time you have implement your decision before things get worse.

I wish I could provide some template or chart to give you some rule-of-thumb about fact-gathering and then making decisions, but to be honest there isn't any. **(See CHART 12-4)** This is not a chart of reality. It cannot really be implemented or referred to. It provides some "ranges of accuracy" for decisions based on having a fraction of all the facts, but it is all subjective.

Decisions are a judgment call that you have to make in real-time, every time. The only absolute we could adopt from this chart is you'll never have 100% of the facts in front of you.

That's why presidents and chairmen of organizations get paid "the big bucks." They have to make tough decisions on many things without having all the facts. Sometimes you have to make a very quick decision and you may only have 5% of the facts in front of you. Will you always be right? No, but not taking any action quickly could be more detrimental than taking some immediate action based on the facts you have.

You have to believe that you can make a good decision once you have adequate facts. That range or percentage or other measure of facts is based on you, your experiences and the matter at hand. You cannot let someone else second-guess you after the fact because then you lose your command and authority. As they say, hindsight is always 20/20. When you are faced with a critical decision in a real-time basis, you must act fast, and think faster.

CHART 12-4: STRUCTURED CHART FOR FACT GATHERING & DECISIONS (ARTIFICIAL – AND NOT REAL FOR REAL WORLD APPLICATION)

FACTS GATHERED	ACTION	OUTCOME	% ACCU-RACY
10%	*KEEP GATHERING*	NO DECISION, MORE TIME EXPENDED	Not applicable
20%	*KEEP GATHERING*	NO DECISION, MORE TIME EXPENDED	Not applicable
40%	*MAKE A DECISION*	DECISION MADE	50-60% accuracy
Over 50%	*MAKE A DECISION*	DECISION MADE	65-80% accuracy
100%	*IMPOSSIBLE TO REACH*	NEVER HAPPEN	Not applicable

Leaders in the political realm (those in local, county and state governments) also need to understand technology, its applications, and how to prioritize projects to get the global *Platform for Commerce* in place. They face the same challenges as their corporate counterparts.

The first Mayor Daley of Chicago, Mayor Richard J. Daley, had big visions of Chicago when he was mayor. He wanted the biggest and the best, from the airport to the water purification plant. He may have had his quirks but he stood by his strategic direction for the city.

When someone criticized his plan, his immediate response was, *"I have MY plan. Where is YOUR plan?"* That usually got the critic pretty quiet.

CARLINI-ISM: *"Don't shoot a challenging salvo across my bow and then cry when I fire back a 44-gun broadside sinking your little ship."*

So Where Is Your Plan?

Today is no different. Have a plan. Be ready to execute it and follow through. Is it the best plan? Let's go back to Patton for a really good answer: A good plan violently executed now, is better than a perfect plan executed next week.

Municipal leaders need to understand technology and how to apply it to government operations. They also need to be able to address challenges about insuring the infrastructure is supportive to next-generation business campuses and regional sustainability. Having next-generation Intelligent Business Campuses (IBCs) which create real jobs as part of a community can only add to its viability, its tax base, and its long-term regional economic sustainability.

Most leaders in municipalities have no idea on how important having a good **Platform for Commerce** is for potential businesses. They must learn fast if they want their areas and regions to sustain regional viability.

Under Richard M. Daley, I was the Consultant to the Mayor's Office for the conceptual planning and design of the Emergency Communications 911 Call Center in the mid-1990s. This was another broad-based project incorporating many real-time computer and radio-based systems as well as a 176-mile network of fiber optics connecting 80 police and fire house buildings in Chicago.

The only real directive he gave was, *"Build a showcase."* And that was what we built!

The Chicago 911 Center was way ahead of its time with dual fiber optic feeds to two central offices as well as the largest non-network carrier fiber optic network connecting over 80 buildings on 176 miles of redundant fiber optics using SONET (Synchronous Optical Network) protocol. At the time, no corporation was looking at campus-based fiber optic services let alone crosstown fiber optic connectivity.

With its back-up center being used for non-emergency 311 calls, it was way ahead of other major cities including New York and Los Angeles. In comparison, New York did not institute a 311 non-emergency service until 2010. A full fifteen years after the Chicago 911 and 311 centers cutover in 1995.

AVOID COOKIE-CUTTER APPROACHES

Throughout the years, many salespeople selling technology sold from a "if it works over there, it should work here for you" strategy which falls short on many systems that had to be customized in order to provide maximum performance. Cross pollinization of technology is great when you can do it, but in reality, most systems and applications need specific tweaking and customization for each customer.

In developing supportive infrastructure, real estate companies need to understand what they need to provide in order to lock in potential corporate tenants for their buildings and business campuses. They also have to work with local political leaders to attain those goals in improving the supporting infrastructure provided by the municipality.

Corporate facilities need to be located on a solid platform for commerce. That platform must include redundant network services from different network carriers as well as redundant power supplies, preferably from two separate sources if it is to compete with other global business entities.

CARLINI-ISM: *"Corporate facilities must support the corporate goals and objectives."*

Again, new rules-of-thumb need to be determined and applied by the local and regional political structure as well. Issues like zoning, tax incentives, building guidelines and permits all have to be reviewed to insure they are up-to-date, relevant, and not a burden or obstacle for developing property and economic development. As pointed out earlier, *"You cannot attack 21st century challenges with 20st century solutions."*

CHAPTER REVIEW

CARLINI-ISMS

"You manage resources, but you lead people."

"A leader can be a manager, but a manager may not be a leader."

"If an organization fails to recognize talent, it starts to lose it to the competition."

"Only performance is reality. Everything else is just glossy rhetoric and slick slogans."

"Empowerment? Spread the responsibilities, but then, spread the recognition and the rewards as well."

"You don't get superman by offering Jimmy Olson wages. Serious people get serious money."

"It's time to be politically accurate and NOT politically correct."

"Take action. Let someone else be patient. Waiting around does not get you anywhere."

"Don't shoot a challenging salvo across my bow and then cry when I fire back a 44-gun broadside sinking your little ship."

"Corporate facilities must support the corporate goals and objectives."

$[13]$

REGIONAL SUSTAINABILITY

"How do you insure future regional viability?"

"Future viability tomorrow depends on building a solid infrastructure today." - JAMES CARLINI

E veryone involved in politics, economic development, community development, and real estate development needs to make a commitment to regional viability and sustainability. Government agencies must understand they cannot just "raise taxes" in order to maintain a region. They must raise and maintain the quality of the infrastructure in order to attract and maintain corporate facilities which will provide jobs, careers, and a viable tax base. These are all positive elements that generate the regional economy.

We have already established the **Platform for Commerce** needs to provide a solid foundation for businesses and other institutions to be built upon it. Having a wide, diverse amount of businesses built upon that platform adds to the local and regional tax base.

CARLINI-ISM: *"If the infrastructure is built right, a region can attract and maintain new businesses. Businesses and other institutions add to the tax base and tax rates will actually go down, making the region an even more attractive area to locate a business."*

Having a strong tax base means you can actually lower taxes and create an even more desirable business climate for organizations to locate within a region. This has been proven with a comparison of two Iowa towns (Cedar Falls and Waterloo) where one added network infrastructure almost two decades ago and because of it, attracted more businesses and actually lowered taxes. Their revenues increased because Cedar Falls had a larger, diverse tax base.

(Cedar Falls vs. Waterloo, IA -http://www.baller.com/pdfs/cedar-falls_white_paper.pdf)

Fresh perspectives focusing on adding technology and new elements of infrastructure must be paid more than lip-service. Fresh perspectives on policy must be taken from the conceptual stage through the stages of planning, design, implementation, and ongoing operations with all the stakeholders in the community.

"Next-generation business parks" which dictate a different approach in local economic development and municipal infrastructure planning are becoming one of several vehicles for developing success in regional sustainability.

The local infrastructure (the ***Platform for Commerce)*** needs to be able to support these next-generation business parks. These parks are tax revenue generators. They are local job generators. They are a way to sustain regional viability.

To implement real-time, communications based, information systems to sustain a competitive lead within a business enterprise is not a one shot deal. It's a continuous process. This also applies to applying network infrastructure to regions in general.

The quality concept of "best practices" has to be eliminated or at least, re-defined, when it comes to government policies and procedures. Remember, "Best Practices" are a moving target. What was state-of-the-art last year, may be obsolete this year. Organizations clinging to their "best practices" are probably clinging to obsolete ways of work and trends as well as infrastructure initiatives and strategic directives.

Taking a management approach of using a total continuous improvement (TCI) methodology on policies and procedures is a much more practical tactic in achieving and sustaining regional viability.

DYING CITIES AND TOWNS

A whole book could be written about dying cities and towns, but let me point out the issues related to infrastructure and insuring a solid *Platform for Commerce*.

As discussed earlier, trade routes have become electronic and critical infrastructure encompasses a whole plethora of new layers which must be discussed and understood to be covered by risk policies. This does not apply only for large cities; it applies for all cities and villages, big and small.

Small towns are starting to realize if they do not improve their infrastructures, they will continue to erode in both a tax base made up of businesses as well as a population made up of homeowners and families. This includes those towns and areas which consider themselves "destinations" and tourist towns.

I spoke to an entrepreneur from a tourist town in the Colorado Rockies who was concerned his town, Estes Park, was not looking towards any long-term planning or improvements to their infrastructure. He said they were losing population and no businesses were moving in.

This is more typical than what you may think. Small towns must become more concerned about their economic survival and look forward, rather than staying stagnant by clinging to the past with their outdated policies.

Think about people going to resorts and other destinations. They are not leaving their Smartphones and Tablets at home. They are bringing them with on vacation AND expect to be able to use them while being out-of-town. If your town's infrastructure doesn't support them, do you really think they will be coming back next year?

CARLINI-ISM: *"Quaint 'Tourist towns' better provide the intelligent amenities to support affluent tourists, otherwise the tourists will flock to another destination that is more 'with it' for supporting their technology needs."*

The tourist town needed to assess what it had in place and analyze what they needed to do in order to strengthen their infrastructure into a better **Platform for Commerce** which could attract new businesses as well as retain the established ones.

They had a local power company which laid some fiber optic cable down. They had a community hospital with some focus on in-house networks as well as a dying school district and local emergency services. The group was composed of a typical combination of organizational players and town "old-timers" who did not want to change the quaintness of their town.

What they needed to address is how they could work together as organizations and civic groups to pool their resources in order to build a common network which all could utilize. The entrepreneur told me they all seemed to be working in different directions.

I told him he should try to get everyone working off the same page in a concerted effort to get the town the proper infrastructure in order to remain viable in the 21st century.

Having some type of common network topology servicing all the factions within the town would not only make it a more viable destination point for tourists, but also a more attractive location for a new group of businesses they could try to recruit to strengthen their tax base beyond the tourism businesses.

CARLINI-ISM: *"Prioritizing projects and getting ideas of who they will impact is an analysis worth spending some time on!"*

Compare the Colorado situation with Jamaica. Jamaica announced they were going to spend over $600,000,000 on upgrading their network infrastructure with fiber optics. They realize their communications infrastructure is critical to maintaining their tourism business.

INFRASTRUCTURE NEGLIGENCE ISSUES

Smartphones are becoming a ubiquitous tool to solve everyday problems from where to buy gas to where to look for a parking space in urban areas. You would think applications (apps) like this would be a positive impact on society. You would think local governments, which on a whole, are not that efficient or effective, would welcome something to alleviate real economic development and infrastructure shortcomings, like parking.

As Smartphone technology becomes more dominant in various parts of our lives, some municipal governments, which cannot accept or understand change, want to fight innovation in order to keep traditional bottlenecks and problems alive and well. Why?

In San Francisco, they are currently embattled with a parking app which provides insight to where there are available parking spaces for those coming into the city. This application cuts into the wasted energy costs of automobiles, not to mention the real problem of a shortage of parking in San Francisco that has not ever been proactively addressed. Motorists looking for a parking spot on average waste twenty minutes of time; and 30% of the overall traffic congestion is created by these motorists looking for a parking space.

SO WHERE IS ALL THE ENLIGHTENED LEADERS?

Why is there such a negative reaction? Is it because the city did not come up with this idea? A lot of classic negative reactions come to mind. We have all seen and heard these classic responses. **(See CHART 13-1)**

When you review the arguments and you look at the problems which have never been addressed, you realize how lame the politicians are to stifle this positive innovation.

NEWSFLASH: In the 21st century, cities need to focus on getting problems solved.

You would think with all the ecology-focused people in San Francisco, this type of app would be heralded as a way to cut pollution, cut wasted energy costs, and make the city more people-friendly by addressing a real infrastructure problem: not enough parking.

CHART 13-1: TYPICAL REACTIONS AND THEIR REAL MEANING

COMMENT	REAL MEANING
"It wasn't invented here."	*Therefore, it is not any good.*
"We will not condone this."	*We are not making any money off of it.*
"This may be innovative, but we need to stop it because it might make things more efficient."	*We don't need efficiencies; we need more bureaucratic overhead to justify our jobs.*
"We don't understand this, so this is bad."	*Even if it helps with a real problem that we have neglected for years to address, we refuse to let it be used to alleviate the issues.*

The solution is not costing any money to San Francisco. Why should they be opposed to a third party with an innovative parking app which shows people where they can find a parking space? (Potentially, this app would cut traffic congestion depending on how many people use it. 10%? 20%?)

Is it because someone else might be making money on their years of neglect of the problem? That would be a good countersuit to their "cease and desist" lawsuit against the app. The countersuit should read "After all the years of not doing anything, here is a lawsuit on infrastructure negligence."

Infrastructure negligence? Wow, sounds like a winner for a new segment of lawsuits. Should the negligence be remedied by fines or jail time by those who committed the negligence?

All of this relates to the **Platform for Commerce**. If a municipality or region fails to build a solid platform for commerce (infrastructure), then its government has failed to maximize the economic development for the region. Shouldn't someone be held responsible for their negligence?

CARLINI-ISM: *"Any government agency impeding the innovation and solution to infrastructure problems (which are impacting regional economic development) should be held liable for infrastructure negligence."*

REVIVING DYING TOWNS

As I was finishing up this book, I went on a spring tour of Savanna, Illinois and neighboring towns along the Mississippi River with a car club I belong to. The tour was interesting because I saw another example of a dying town.

At one time, Savanna, Illinois was a town which had a diverse economy employing many with the railroads as well as an Army Depot. The Savanna Army Depot, after being used since 1918 closed down in 2000 as part of a federal overall base realignment and closure process. There was a job loss of 450 full-time jobs. The Depot utilized 13,000 acres of land, but only part of the parcel (3,000 acres) is currently available for re-development.

The land may not be able to be re-purposed or given up for a new commercial initiative because of unexploded ordinance and other environmental issues. The land may never be totally reclaimed and re-used.

Amenities listed for the land include broadband, but no speed capacity or carriers are listed. More importantly, no natural gas utility serves the property.

They need to add more definition to their description of the land as well as improve the actual infrastructure if they really want to attract and maintain new viable businesses.

The railroad shut down its facility as did the Army Depot and the town's economy took a nosedive. The town lost 75% of its population.

Within the downtown area of Savanna, there are several storefronts for sale among several bars as well as a county museum which is under renovation. The bars include the Iron Horse Social Club, owned by a motorcycle-loving entrepreneur who owns several of the storefront bars. He is focused on trying to revitalize the downtown area by trying to create another Sturgis by attracting bikers and other tourists to the area. Jerry told me some of his plans and he really wants to make things work. Like so many other creative people, he has the creative ideas, but not as much in capital. He has one vision of where the town should go.

Another group which promotes another vision is the one which runs the county museum just down the street. They have a small museum which displays a large and diversified number of both Union and Confederate uniforms on one floor of their museum. They want to attract a different crowd. Another faction is the owner who owns a 62-room castle only a few blocks away. He has a different vision of bringing people into the various attractions.

What all these diverse stakeholders need to do is sit down and combine all their ideas and efforts to get the momentum moving forward for the whole region. They should try to sit down with the local government and community development people to analyze what they can build together, instead of trying to go in several directions at once with limited funding. This is not an isolated case. If anything, this is a common issue across many small to large municipalities. What do we do to survive? What do we do to thrive? How do we set a strategic direction? Sounds simple to ask, but many are not asking the right questions.

CARLINI-ISM: *"You cannot get the right answers if the right questions are not being asked."*

In another part of Illinois, the county cannot seem to fund different services for its constituents so what they have done is load up fees as part of the "court costs" for traffic court. Talk about living in the County of the Sheriff of Nottingham, this county depends on revenues generated from additional fees tacked on to misdemeanor traffic tickets which have no tie-in or correlation to the violation itself. **(See CHART 13-2)**

When you look through the fines tacked on the court costs, there is absolutely no correlation to a speeding ticket or other moving violation.

Mental Health Court fee? Yes, I guess you have to be crazy to buy off on these "tacked on" charges. Drug court? No drugs were involved. Why is this "tacked on?" State Police Fee? They weren't involved. No violent crime and no children were involved, yet there is another $83.00 tacked on for all these services.

Maybe everyone showing up should charge the court a "guest appearance fee?"

CHART 13-2: THE SHERIFF OF NOTTINGHAM COUNTY (IN ILLINOIS) BREAKDOWN OF "EXTRA COSTS AND FEES"

DESCRIPTION	ADDITIONAL COURT COST
CLERK PROCESSING FEE	$5.00
CIRCUIT COURT FEE	$15.00
COURT AUTOMATION FEE	$15.00
ELECTRONIC CITATION – VILLAGE ORIGINATING TICKET	$15.00
MENTAL HEALTH COURT FINE	$10.00
DRUG COURT FINE	$5.00
COURT FEE	$5.00
COURT SECURITY FEE	$25.00
ATTY FEE – VILLAGE ORIGINATING TICKET	$25.00
STATES ATTORNEY AUTOMA-TION	$2.00
CHILD ADVOCACY CENTER FINE	$13.00
VIOLENT CRIME ASST. FUND	$50.00
STATE POLICE OP ASSISTANCE	$15.00
TOTAL	$205.00

Looking at the court cost breakdown, which by the way, does not include the actual fine for the statute that was broken, shows a clear lack of managing budgets and trying to balance those budgets with horse-and-buggy fines instead of solid principles of economic development. Based on who gets a cut of the ticket, the municipalities as well as the county are guilty of mismanagement of their budgets when you look at how they all need to get an "extra piece" of revenue so they can pay for something they incorrectly budgeted within their jurisdiction. And people wonder why Illinois is so messed up with budget shortfalls from state budgets to individual municipal agency budgets?

Doing Things Right? Understand The Regional Economic Framework

Let's try to visualize what needs to be done and how it all is interrelated. Public policy affects Levels 1 through 5 in the Regional Economic Framework (REF) and must provide a clear, strategic, unified direction identified and understood by all interested parties. **(See CHART 13-3)** Each layer builds upon the layer beneath it.

Natural Resources (**Layer 1**) are what everyone starts out with in their area or region. The topography is a given. If you are in the desert, you can't change that to rolling hills. You need to build upon the natural land which is what every city and village has already done for the last one or two or more centuries. Every municipality should know at this point what their strengths and weaknesses are, given the topography.

Infrastructure (**Layer 2**) is where the **Platform of Commerce** gets built. What do you have in place? What should you add on? What do you or a group of investors (stakeholders) have to invest in the infrastructure?

As already established, infrastructure has been around for 5,000 years. Chances are, yours needs to be upgraded.

Man-made Resources (**Layer 3**) are where all the businesses and other institutions are built. These are structures that can be built, removed, reconfigured, or modified on top of the ***Platform for Commerce.***

Jobs (**Layer 4**) are predicated on what types of businesses and organizations have a presence in the local region.

With broadband connectivity, you can also extend jobs into areas that physically don't have the businesses and their facilities present. People who are "telecommuting" to jobs are a growing segment in the workforce.

The Regional Economy (**Layer 5**) is the sum of all the underlying layers. If you have a strong foundation, you will positively impact the regional viability of the area. If the ***Platform for Commerce*** is weak or missing critical layers, what is built upon it will suffer. The end result is that the regional economy is not good and may not even be sustainable.

Having a good broadband layer within the ***Platform for Commerce*** can actually add jobs into a local economy or an area where you physically don't have to build the building to house them.

CHART 13-3: THE REGIONAL ECONOMIC FRAMEWORK

LAYER	LEVEL	DESCRIP-TION
5	*THE REGIONAL ECONOMY*	"Regional economic climate" – everything added up.
4	*JOBS*	"The collection and re-circulation of salaries"
3	*MAN-MADE RESOURCES*	Schools, Hospitals, Factories, Office Parks
2	*INFRASTRUCTURE*	"The *Platform for Commerce*": This is what businesses are built upon.
1	*NATURAL RESOURCES*	"What the region offers in natural resources."

PROJECT FUNDING PRIORITIZATION

With the distribution of federal Stimulus money several years ago, requests for funding of various layers of infrastructure and road projects were solicited and for the most part, not clearly reviewed, analyzed, and prioritized for their impact to the community in most areas and legislative districts within all the states.

In one legislative district in Illinois, a methodology was developed and put into practice. I worked with State Senator Mike Noland to develop an objective methodology for him to review projects he was reviewing to determine their viability for funding.

It was a critical step in defining and prioritizing what projects were to be funded. The need for providing a structured approach was considered important in these days of "transparency" and accountability.

With two different funding mechanisms which had to be utilized for maximum benefit (the Federal Stimulus Package and the state's Capital Funding Program), a structured approach was needed and should have been adopted by every legislative office.

It should be noted, there were also other factors and selection criteria which could not be easily structured into this evaluation process. As an initial first step, this process gave a much clearer picture in sorting out over 100 projects which were presented for funding. Some simple questions had to be asked in order to start the evaluation process of this mix of road projects, interchanges, municipal facilities and other neighborhood beautification projects (parks, centers, and other programs).

This model provides a structured approach to analyzing where limited funding can best be applied and utilized for the greater good of the legislative district and the state, instead of a purely subjective approach that most politicians use and abuse.

In these financial times, keeping the same level of funding for every agency may be considered a non-attainable accomplishment and in many cases, totally impossible. No agency should expect an automatic increase in funding.

If anything, agencies should start preparing themselves to keep operating when a budget cut becomes a reality. If you are a director of one of these agencies, can you cut 10% out of your budget?

WHO BENEFITS FROM THIS PROJECT?

This seems to be a simple question to ask, but many forget to ask it. First, a project should be defined as beneficial using the **ICARE** © Model, a five-level approach to identifying projects and their impact as developed by the author (James Carlini).

You need to define who will benefit if the project gets funded and undertaken by the task group. **(See CHART 13-4)**

This first evaluation of the project would be to designate where the project's footprint for benefits are realized. Does it cover the following?

an individual group **(I)**,

a single community or municipality **(C),**

a group or area of municipalities **(A),**

a full legislative district comprising of multiple municipalities **(R1 or R2)**, or

the entire state **(E)?**

CHART 13-4: THE ICARE MODEL

LEVEL	ABBREVIATION	DEFINITION
1	I	Individual organization or group
2	C	Community (single municipality or township)
3	A	Area (several municipalities and/or townships)
4	R	Region (the full legislative district (R1) or multi-regional districts (R2) where the project overlaps two or more legislative districts)
5	E	Everyone in the State (statewide)

How does this fit in the overall scheme of things? The impact of the ICARE project funding fits within the 2nd through 4th layers of the five-level Regional Economic Framework. **(See CHART 13-5)**

The next prioritizing step was defining the project cost as either a **MEGA, SIGNIFICANT, MAJOR, LARGE**, or **SMALL** project cost.

CHART 13-5: HOW EVERYTHING FITS TOGETHER

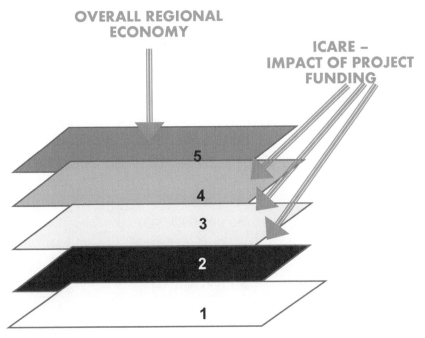

OVERALL REGIONAL ECONOMY

ICARE – IMPACT OF PROJECT FUNDING

5

4

3

2

1

Five-level REGIONAL ECONOMIC FRAMEWORK

Total or partial funding could be at the discretion of the executive or committee based on various factors: access to other funding, internal funding, revenues generated (tolls, fees, business licenses) or other alternatives. **(See CHART 13-6)**

Once the initial benefit analysis was established, a tertiary criteria could be applied to prioritize to each project ranking within that level:

- **CRITICAL**
- **NECESSARY**
- **OPTIMAL**

CHART 13-6: THE ICARE © MODEL – PROJECT SCALE

SIZE OF PROJECT	DOLLAR AMOUNT	ABBREVI-ATION
MEGA	(over $10,000,000USD in funding costs)	MEGA
SIGNIFI-CANT	(over $5,000,000 to $10,000,000USD)	SIG
MAJOR	(over $1,000,000 to $5,000,000USD)	MAJ
LARGE	(over $250,000 to $1,000,000USD)	LRG
SMALL	($250,000USD or less in funding)	SMA

This three-level ranking approach provides a second sorting refinement to prioritize road and infrastructure projects. It provides a realistic gauge as to what should be best for each category that focuses on individual organizations, communities, areas (multiple communities) and regions (full legislative districts).

Those factors would be applied after the objective sorting process was completed including the use of the Critical, Necessary and Optimal Categories. **(See CHART 13-7)**

CHART 13-7: CRITERIA FOR SERVICES

LEVEL	DESCRIPTION
CRITICAL	Provide critical services that should not be cut. Examples would include, but not be limited to: Public Safety (first responders), Public Health, infrastructure (critical platform for commerce and economic development), critical bridge or interchange.
NECESSARY	Provide necessary services. Examples would include, but not be limited to: Schools, Community Colleges, other public works.
OPTIMAL	New social, educational program or benefit, expansion of existing public services, park expansion, bike paths, or anything that is not considered **Critical** or **Necessary** services.

The distillation of each project's priority can start with this ICARE ©Matrix. As stated earlier, this would not be the total review or final review performed on each project, but would be the initial steps in identifying, categorizing, and prioritizing projects using a structured, objective approach as a foundation for selection criteria instead of something less positive or totally subjective like the whim of a politician or special interest group.

Sample chart of Project Funding Matrix **(See CHART 13-8)** where each project submitted would be categorized into the above matrix to begin to sort out and prioritize all the endeavors of various organizations, agencies, and municipal entities.

CHART 13-8: PROJECT FUNDING MATRIX (Sample)

	CRITICAL			NECESSARY			OPTIONAL		
	MEG A	SIG-MAJ LARG	SM A	MEG A	SIG-MAJ LARG	SM A	MEG A	SIG-MAJ LARG	SM A
INDIVID-UAL	PRO-JECT 1, 2			PRO-JECT 4			PRO-JECT 8		
COMMU-NITY					PROJECT 6				
AREA	PRO-JECT 3							PRO-JECT 7	
REGION				PRO-JECT 5					
EVERY-ONE									

Source: James Carlini

Just taking the time to categorize and rank all the projects is a good exercise to understanding what is being proposed and how it will affect the region. It is time well-spent.

CAN-DO APPROACH TO PRIORITIZING PROJECTS

A four-level variation on the three-level priority criteria can also be used. Instead of:

- **CRITICAL**
- **NECESSARY**
- **OPTIMAL**

A four-level criteria can be applied to prioritize to each project ranking within that level. **(See CHART 13-9)**

CHART 13-9: CAN-DO PROJECT APPROACH

CRITICAL
ALWAYS NECESSARY
DESIRABLE
OPTIMAL

This four-level ranking approach (CAN-DO) provides a second sorting refinement to prioritize projects and afford a realistic gauge as to what should be best for each category that focuses on individual organizations, communities, areas (multiple communities) and regions (full legislative districts). Those factors would be applied after the objective sorting process was completed including the use of the CAN-DO Approach - Critical, Always Necessary, Desirable and Optimal Categories. **(See CHART 13-10)**

CARLINI-ISM: *"Prioritizing projects and showing the methodology used to compare them provides a clear answer to whether or not money was well-spent."*

This provides an alternative structured approach to analyzing projects into four levels where limited funding can best be applied and utilized for the greater good of the legislative district and the state. In these financial times, keeping the same level of funding for every agency may be considered a great accomplishment. More likely, many agencies may be facing cutbacks in programs as well as funding. No one should expect an automatic increase in funding.

In most cases, I do not think many government agencies do a good job in prioritizing projects and/or budgets for various departments. If they went through this type of exercise for every department or agency, I believe they would spend money more wisely and would positively affect more people that just going through and picking the projects on more of a political process than an objective, prioritized project funding approach.

When you take this approach to deciding what should get funded, you automatically create a very visual record of how you came to your budgeting conclusions. It is simpler to explain when you have a structured format you can refer to.

CHART 13-10: CAN-DO APPROACH

PRIORITY	LEVEL	DEFINED AS
1	CRITICAL	Provide critical services that should not be cut. Examples would include: Public Safety, Public Health, infrastructure (platform for commerce and economic development)
2	ALWAYS NECESSARY	Provide necessary services. Examples would include, but not be limited to: Schools, Community Colleges, public works.
3	DESIRABLE	Something that would be more broadly advantageous in its applications and coverage.
4	OPTIMAL	New social, educational program or benefit, expansion of existing public services, anything not considered Critical or Necessary services.

LEGISLATIVE IMPACT

Another measure which is more subjective is to define IMPACT. How do we first define societal/political IMPACT and then measure it, prioritize it, or at least give it some type of value? IMPACT is also a multi-faceted measure. **(See CHART 13-11)**

CHART 13-11: LEGISLATIVE DISTRICT IMPACT

POLITICAL	Obvious – how many votes does this directly impact?
ECONOMIC	Does it create jobs, if so how many, for how long and what type?
ENVIRONMENTAL	Positive, negative, or neutral for the environment?
REVENUE	Any revenues involved? Potential sales tax or user fees?
OPERATIONAL	Getting Government better, more efficient, cost reduction.

A weighted approach to addressing each one of these criteria further defines where a project fits within a project funding list.

The vast majority of different state and congressional districts used a much less structured and purely subjective approach, but it is hoped that in future distributions of project money or annual project funding reviews, a more structured framework and objective prioritization process could be used.

When Chicago looked at hosting the Olympics, it was going to add a lot of new infrastructure in order to support the logistical operations of the Olympics.

If they used the **Platform for Commerce** approach, they could have easily developed plans to address each layer of infrastructure and what it needed as improvements and upgrades to its current state. In addition, it could also determine what elements of the infrastructure they wanted to make as permanent additions and what elements would only be temporarily augmented in their construction and usage. **(See CHART 13-12)**

EXAMPLES:

1)   In planning the Olympics venue, a total assessment of what is in place and what was needed to be upgraded permanently (P) and what was needed to be augmented (A) just for that event would have been needed to be performed on every layer of infrastructure. A matrix like this might have been created.

2)   Prioritizing projects need to have certain sorting processes to identify more critical endeavors than those which can be implemented at a later time.

By using these concepts and frameworks in analyzing projects, municipalities and state agencies can better determine objectively what projects should be funded first and given a higher priority which distills the process into a more "must-have versus hoped-for" lists. Only then, will projects be implemented for their impact on the economic sustainability of the region as well as a more objective, prioritized sequence of choices that they should be selected by for funding, than on emotional or uninformed decisions.

CHART 13-12: NEW METRICS TO REVIEW ALTERNATIVES

PROJECT	INFRASTRUC-TURE LAYER	PERMA-NENCY
MASS TRANSIT (LIGHT RAIL)	3	P
MASS TRANSIT (BUSES)	2	A
MASS TRANSIT (BOATS)	1	A or P

CARLINI-ISM:　　　　*"If a new technology is proven in a major application, it is hard not to try it somewhere else. The risk is minimized."*

CHAPTER REVIEW

CARLINI-ISMS

"If the infrastructure is built right, a region can attract and maintain new businesses. Businesses add to the tax base and tax rates will actually go down, making it an even more attractive area to locate a business."

"'Quaint Tourist towns' better have the intelligent amenities to support affluent tourists, otherwise the tourists will flock to another destination that is more 'with it' for supporting their technology needs."

"Prioritizing projects and getting ideas of who they will impact is an analysis worth spending some time on!"

"Any government agency impeding the innovation and solution to infrastructure problems (which are impacting regional economic development) should be held liable for infrastructure negligence."

"You cannot get the right answers if the right questions are not being asked."

"Prioritizing projects and showing the methodology used to compare them provides a clear answer to whether or not money was well-spent."

"If a new technology is proven in a major application, it is hard not to try it somewhere else. The risk is minimized."

[14]

CASE STUDY -IBC

"The need to innovate or die." – **JACK TENISON**, Executive Director, **DNTP**

The DuPage National Technology Park (DNTP), which was later renamed the DuPage Business Center (DBC), had been originally developed as a regional platform to attract and maintain sophisticated, corporate tenants who would provide new job opportunities within Illinois. The DBC is a good example of a next-generation Intelligent Business Campus (IBC) because of its focus on providing elements of intelligent infrastructure upfront before corporate tenants actually decided to move in.

The location was considered to be a prime area to attract sophisticated facilities for various corporations as an 800-acre business campus next to a regional airport. It also had a large data center built in it as well.

Some of the key intelligent infrastructure attributes this IBC offered were multiple network carriers providing high-speed access to different networks. The infrastructure provided redundancy in both network communications and power sources. These intelligent amenities addressed and guaranteed the support of mission critical applications that a sophisticated corporate tenant might have.

WHAT WENT RIGHT WITH THE DBC?

The things which went right with this project included the backing of then-Speaker of the House Dennis Hastert, as well as all the surrounding towns within the area of the 800-acre park. This new technology park became the example of a "next-generation" business park for others to emulate. The goal of being a catalyst for job creation and attracting new companies into Illinois was a big reason for the next-generation park to be built. The technology and business campus idea was heralded as the best use of this land for local and regional economic development.

One of the big breakthroughs I saw in this project, was the ability to get everyone moving forward together and focusing on the right ways to streamline typical municipal and state bureaucracies' zoning issues. Red tape and slow government agency responses, which would typically hinder a project's completion, was to be eliminated wherever possible. If you have ever served as an elected official, you understand how bureaucratic policies and little items being debated can explode into time-wasting arguments filling meeting agendas as well as monthly discussions at various sub-committees. All of this wastes time.

CARLINI-ISM: *"Where is everything going? Time is not money – time cannot be replaced!"*

This united government agency approach should not be taken lightly as an idea to be emulated by other regions. City, county, and state resources working together should be a given and not a hoped for. Unfortunately, many projects in many states do not have this joint commitment.

CARLINI-ISM: *"Rule for government agencies: If we are all in the same boat, let's make sure we all have an oar in the water and are pulling together."*

Streamlining the process and making sure there are no bureaucratic stumbling blocks can make the difference of getting a good project off the ground successfully or losing it to someone who is more adept at being "business-friendly" and knowing where the "bureaucratic boat" has to go.

State and federal grants to help build out the initial infrastructure were awarded to the DBC project. In projects prior to this, intelligent infrastructure (power and network communications) would be left to plan and build out after corporate tenants moved in.

The amount of the grants awarded totaled $40,000,000 which sounds like a lot, but when you compare the support to other countries' funding support of next-generation industrial park and building complexes, it is not very substantial.

Compare the project with CyberPort in Hong Kong, where the government put up the money to buy the land for the complex.

They provided $1,000,000,000 to purchase the land. That is the state of the competition out there.

Other state-supported business campuses can be seen in Taiwan and in Mainland China. The Asian governments understand the value of becoming a strong competitor in global markets and they are not afraid to subsidize some of these new real estate business campus and industrial park endeavors.

Another key area the DBC project focused on was the availability of more than one network carrier for connectivity into the network infrastructure layer. Besides having several network carriers terminate into the Park, access to other broadband infrastructure including the National Lambda Rail (NLR) which is a specialized, high-speed network for research facilities and major universities, was made available.

Having in place power and network communications providers gave the DuPage Business Center more of a finished look to prospective corporate tenants. Corporate site selection committees are looking for places where there are no questions when it comes to whether or not it is going to have these intelligent amenities. If there is a question or a doubt, they just move on.

I reviewed all the network communications facilities being brought into the DBC as well as the conduit issues within it. I also reviewed some of the inter-municipality agreements concerning maintenance of some of the various conduits providing access for fiber optics.

Another network bottleneck was avoided by not going with the suggestion to "just let the phone company come in and build out their network infrastructure into the complex."

The problem is once they build the conduits into the campus, they have locked in all the access as to what other carriers can come into the complex and control what they can charge them for access to the corporate tenants within the campus. They have effectively painted the property owner out of the picture.

CARLINI-ISM: *"Avoid offers from any incumbent phone company to 'provide the business park's network infrastructure.' If you do, you just choked off connectivity."*

Some would argue this and ask, *"Why not let the phone company put in the conduits for free? It's a great idea and will save us money."* The problem is you have just given them the complete control of who comes through the conduits. They are not doing this to save you money, they are doing it to save their market share.

They will take a very active role in making sure anyone in the complex gets their service first. It defeats the whole idea of having broad access to network services. Remember, it's "Location, Location, Connectivity", not "Location, Location, Restricted Access."

In addition, I reviewed comparable industrial parks in Asia as part of my research. **(See CHART 14-1)**

I had a chance to meet with the Taiwanese consulate and found we were not really on the cutting-edge as much as in a catch-up mode where we did not even know a race was going on.

In some of my research, I found there were government initiatives in Taiwan dating back to 1999 to establish these next-generation Intelligent Industrial Parks (IIPs) in various areas of Taiwan.

They were all focused in taking products from R&D to commercialization in various high-tech industry sectors.

CHART 14-1 NEXT-GENERATION BUSINESS PARKS

UNITED STATES	ASIA
Intelligent Business Campus	Intelligent Industrial Park
IBC	IIP

Further investigation uncovered mainland China was following the Taiwanese in building specialized IIPs in different provinces of China as well. Hong Kong had CyberPort, but more importantly, different provinces like Hunan, had their Intelligent Industrial Park complex as well. Again, each one focusing on some industrial products.

All of this was eye-opening and reinforced our need to get moving in this direction. Our major trading partners were already on the same track in changing traditional real estate into high-tech real estate supporting complex businesses competing in various global markets.

In several presentations at various national conferences between 2008 and 2013 when I gave the DuPage Business Center as an example of an Intelligent Business Campus, I commented on our national need to become more competitive and focus on other countries already making the strategic move of having next-generation business parks, with very high-speed network access, supporting various industries.

We need to stop dragging our feet and apply ourselves to compete. We need to understand real estate requires intelligent amenities to support businesses in the global 21st century marketplace.

Moving all our network infrastructure forward to be able to compete globally, should be a national, strategic priority and not just a regional exception to the norm.

One of the other ideas was to re-define the traditional name of this next-generation business park to give it a more descriptive name to make it stand out from the traditional parks of the past. I wrote a white paper defining the **Intelligent Business Campuses: Keys to Future Economic Development**. Within the white paper, I discussed the new areas of having power and network communications defined in the upfront "Master Planning" of a development, instead of as an afterthought when corporate tenants had already moved in.

Redefining this project was important because it should not be lumped in with all the traditional real estate business parks and industrial parks of the past. A whole new set of terminology and definitions have to be used to describe this paradigm shift in real estate. This terminology has been described throughout this book.

What Went Wrong With The DBC?

Let's take a quick post-implementation review of things since the DBC campus opened. It is one thing to talk about future use and future tenants, it is another thing to discuss what is happening today and what opportunities never revealed themselves.

Blame can be put on one group or another, but the bottom line is they are not attracting the caliber of corporate tenant which the DuPage Business Center was aiming for.

One issue pointed out to me was the option of buying the land for the corporate facility. Some companies wanted the option to buy and the DBC declined to sell land. They were offering a lease-only agreement and evidently some companies prefer buying the land outright before building their corporate facility.

Those who are managing the DBC campus just don't seem to get it. I am not sure what their issues are except they don't realize what they have and are not packaging the park's capabilities correctly to sell in today's environment. Again, it may be the managerial disconnect of having people stuck with a 1950s sales mentality trying to sell space in a 21st century market. If they are not using the new terminology defining the park and its intelligent infrastructure attributes, chances are, they are not selling its strongest qualities.

CARLINI-ISM: *"New terminology and definitions need to be learned if you want to be able to design, build, sell, and operate real estate in the 21st century."*

Another reason the park is not filling up with corporate tenants is the actual business climate in Illinois is not the greatest. In fact, it is one of the worst when you compare it to other states trying to attract and maintain new corporate facilities. Real job creation is a high priority.

Companies looking for new locations have their pick across the United States because every state is trying to secure a better economic climate.

States getting new corporate facilities in, which in turn, create jobs, are competing on many levels both domestically and in some cases, globally. This is of high importance on every Governor's to-do list:

Attract companies to locate their corporate facilities in the state to create jobs.

I called those who took over the management of the DuPage Business Center, but I never received a return phone call. I emailed them as well but never heard anything. The mere fact they didn't return a call tells me they may have missed returning other people's calls. Returning a call seems like such a basic thing to do but maybe in today's business environment, they can pick and choose the calls they want to return. Most real estate and property management firms do not have that luxury.

CARLINI-ISM: *"Return calls. Nothing is more unprofessional than people who do not return calls. The call you don't return, is the call which could have made you $1,000,000."*

OTHERS GAINED INSIGHTS FROM THE PROJECT

The DuPage Business Center was not a complete failure. Others gained clear insights from the way it was built out in its initial stages and design.

Working with another developer in Illinois, the ideas of adding infrastructure upfront was viewed as a good idea to keep the 120-acre Terra Business Park flexible when it came to attracting new technology-dependent companies.

At Terra Business Park, in East Dundee, Illinois, all the conduits for fiber and for power were added into the ground before the streets into the complex were completed. The incremental cost of putting the conduits in was minimal compared to what it would cost once the ground was built out with both roads and landscaping.

The owner saw there was merit in offering intelligent amenities upfront for corporate tenants, like redundant power sources and multiple access to network carriers. He was impressed by the comparison numbers which focused on buildings that provided these services compared to buildings that did not.

The research done for the DuPage Business Center clearly pointed out offering space is a commodity approach to selling real estate properties whereas offering intelligent amenities, like broadband connectivity, changes the game and eliminates over 90% of the competition which is vying for a blue-chip, corporate tenant.

We also analyzed what it would cost to connect up to a multiple-carrier fiber optic right-of-way on a nearby interstate highway. It was only a couple of miles and the idea of running an underground fiber from the business park to the interstate was considered. This type of capital expense would be a great item to ask local government support on. In many areas where there are TIF districts, this type of project would be a great one to get funding on. Why? It fits the criteria for what the TIF monies are all about – giving developers some help on infrastructure issues in order to insure a higher-caliber development which in turn should provide more jobs and more of a residual impact on regional economic development.

Know The Power Of TIF Money

Working with some real estate developers as well as being an elected official for a term gave me insights as to the benefits of TIF money on projects. Many municipalities have created business zones where TIF money can be used to attract a better development. The monies are to be used for improvements within the infrastructure.

The problem is many people who sit on these boards have no idea as to what needs to be added to the development. Again, it goes back to what their definition of "infrastructure" was and most do not comprehend all the layers as defined in this book. We need to overcome that because as seen in the DBC project, the layers of infrastructure have to include network and power grid connectivity.

What is the best use of money for today's and tomorrow's projects? That should be the question raised by any municipal board focused on developing a TIF fund for a business development area. It seems like an important question to raise and answer, yet many boards are not seeing past old rules-of-thumb and old (incomplete) frameworks of infrastructure.

With the explosion of growth in wireless communication devices along with new apps that offer consumer services, having network connectivity is a critical amenity. A next-generation Intelligent Business Campus must have broadband connectivity for individual (consumer) applications as much as it should have connectivity for corporate applications.

CHAPTER REVIEW

CARLINI-ISMS

"Where is everything going? Time is not money – time cannot be replaced!"

"Rule for government agencies: If we are all in the same boat, let's make sure we all have an oar in the water and are pulling together."

"Avoid offers from any incumbent phone company to 'provide the business park's network infrastructure.' If you do, you just choked off connectivity."

"New terminology and definitions need to be learned if you want to be able to design, build, sell, and operate real estate in the 21st century."

"Return calls. Nothing is more unprofessional than people who do not return calls. The call you don't return, is the call that could have made you $1,000,000."

[15]

CASE STUDY: IREC

"The revival of real estate is facilitated through applying intelligent amenities." – JAMES CARLINI

In the last decade, we have seen the emergence of next-generation Intelligent Business Campuses (IBCs) in the American commercial real estate sector and Intelligent Industrial Parks (IIPs) in other global markets. They have become an integral part in regional economic development and need to be understood as a vital part of the overall platform for global commerce.

We are now witnessing the first applications of broadband technologies, high-capacity wireless networks, Smartphones, digital signage and video applications to retail, entertainment and convention center campuses as new intelligent amenities to draw and maintain consumers in an entertainment/ retail environment. These campuses can be referred to as Intelligent Retail/Entertainment/ Convention center (IREC) Complexes which can be considered another critical platform for commerce in the 21st century.

With more Smartphones being utilized everywhere, should we be redefining Cloud Computing? If not redefining it, at least re-calibrating it to encompass and fit new edge technology which is becoming the device of choice.

Many organizations looking at implementing Cloud Computing should also be looking at BYOD (Bring Your Own Device) concepts which focus on Smartphones and Tablets. In any BYOD environment, the key is to have support that is not focused on any one operating system, but focused on supporting multiple platforms of operating systems (like Android, iOS, BlackBerry, and others)

I have seen where some people do not realize BYOD is "operating system-agnostic" (meaning it should not care what Operating System you are using). The network support should address all the devices you have on your network (or your cloud). If you decide you are only supporting one type of operating systems like iOS (for Apple) or Android (for Samsung, LG, and many others), you may be limiting your success.

THE IMPACT ON RETAIL AND ENTERTAINMENT VENUES

In the past, both retail centers and entertainment venues were not too concerned about providing access to high-speed network services. Today, with the explosive use of Smartphones and other personal digital devices, the need for cellular services to provide more than just voice communications capabilities is a must have, rather than a hoped for in all multi-venue facilities.

New and emerging services like mobile wallet "purchasing at your fingertips" applications for food and drink concessions, team wear, and other "event experience memorabilia" need to be available and capable of handling volumes of impulse purchases.

CARLINI-ISM: *"Let your fingers do the walking, has transformed to, 'Let your fingers do the buying,' with today's Smartphones."*

If done right, the wireless capabilities should create a *"Virtual Resort"* effect over the multiple venues where the attendee (customer) does everything within the confines of the "complex." Think of an electronic umbrella encompassing the real estate parcel.

There is no need to go outside of the *"Virtual Resort"* perimeter to buy things. All purchases, from food concessions and merchandise to dinner and drinks afterwards, are done within the complex's perimeter and within the reach of a Smartphone.

Buying into this concept requires a multi-level sell. Building developers and owners need to buy off on this strategic concept as well as the potential tenants and the municipality itself. If they do, they can create a very powerful economic engine.

A MULTI-LEVEL SELL: CREATING THE "VIRTUAL RESORT" EFFECT

To design one venue, like a stadium or convention center, requires a certain level of expertise to plan and implement a working solution. When you add several different venues (entertainment, restaurants, and retail) together into one complex, the need to understand the interrelationships which are going on and the potentials for cross-selling and cross-marketing as well as the overall changes in network traffic patterns and usage peaks is critical.

What are the benefits to the building owners of building an electronic *"Virtual Resort"*? This intelligent amenity provides a competitive advantage for the complex in enticing prospective tenants by offering added capabilities to insure their business's success as well as providing a more enhanced experience to anyone visiting any part of their multi-venue complex. Instead of space being a commodity, these added intelligent amenities makes the location more valuable to both the owner and the tenant because the business tenant now has more electronic marketing (*eCOUPONS, ePASSES, eDISCOUNTS*) pushing people into his establishment than in a traditional real estate space. For the building owner, having this capability is a distinct competitive advantage which will attract a higher-caliber tenant and create higher occupancy rates.

Ask any building owner or developer if they know of any other strategy to eliminate over 90% of their competition when it comes to enticing tenants to choose their development over others.

In the next phase of implementation, the electronic capturing of consumer demographics can further provide more insights and critical information on what the venue's target market really is. This additional layer of consumer analysis is not available at any traditional real estate complex. Once this is also implemented, there is effectively no competition to the *"Virtual Resort"* effect.

What are the benefits to the tenants? The retail tenants within the complex benefit by having this *"Virtual Resort"* effect capture customers which they may not have gotten in traditional properties that do not offer this electronic mobile cross-connect of businesses, venues and demographics. It ties in with the already established concept: The three most important words in real estate are now, *Location, Location, Connectivity.*

Capturing the lost customer through Smartphone applications like *eCOUPONS* and *eDISCOUNTS* can add a substantial percentage of new business to a venue which would have to spend money in more traditional advertising means. Remember the old adage about traditional strategies? *Advertising is only 50% effective, but how do you choose which 50% to cut out?*

Think about tying together digital signage where you can advertise in one venue the offers and deals going on in real-time at other venues within the "***Virtual Resort***." When walking through the entertainment complex (stadium or resort), you can see "*e-Offers*" being generated by restaurants or other retail shops within the area. People will act on seeing those motivational *eCOUPONS.* You can literally steer them to other venues. Keeping them "on the resort" translates to getting more revenue per customer visit. That benefits the retailers and restaurants as well as the municipality itself which is collecting the sales tax revenues on all the purchases.

What are the benefits to the municipality? This is another very important facet supporting the concept of regional economic sustainability mostly overlooked by local economic development and community development commissions. The municipality also benefits from "*capturing the lost customer*" as it gains more sales tax revenues which may have gone to a neighboring municipality or neighboring state. The pervasive consumer attitude becomes "why go further than what is offered right here as far as restaurants, entertainment, and other retail offerings?" The cities and towns, as well as the local economic regions, need to understand this.

When it comes to municipalities enticing organizations with TIF (Tax Increment Financing) or Business District Development (BDD) money to help make initially locating and developing in a community a more viable decision, the whole area of network infrastructure becomes a strategic initiative to help fund instead of the more traditional ways TIF monies are spent on more conventional improvements like streets, building facades, and parking lots.

Declaring an area as a *"Business Enterprise Zone"* or an area "ready for commercialization" is not as important as actually preparing it to be a *"Business Enterprise Zone."* The infrastructure has to be able to support the demands of a sophisticated event, like a tradeshow or sporting event, just like a commercial building complex needs to address the high-tech demands of corporate tenants.

Having this initial network infrastructure in place, provides an extra layer within the *Platform for Commerce* that this business will build upon. Huge cash giveaways for tax incentives to attract a certain venue or store become a thing of the past.

What are the benefits to the customers and those attending the different venues?

Individual customers and those attending the conventions and entertainment events all have their ubiquitous "edge technology", their Smartphones and Tablets. They want to be able to use them to their full capabilities. In each scenario, the customer wins – but so does the combined organizations that maintains the *"Virtual Resort"*:

-      What better way to spend the day at a convention center. The exhibitors can be contacted via a Smartphone for more marketing materials as well as product information. After the convention, the attendee can stop at a near-by restaurant or bar to cash in on the electronic *eCOUPON* they received earlier on their Smartphone.

-   The attendee comes to the sporting event and can order drinks, food, and other concessions via their Smartphone. They can order team wear and other team-related products all from their seat. Before the games ends, they will be sent some redeemable electronic coupons so that they stop at a near-by bar or restaurant before driving all the way home.

-   A family comes in for dinner at a restaurant and is sent electronic *ePASSES* for an upcoming entertainment event or tradeshow.

All this cross-marketing makes the whole area more viable because with every person coming into the *"Virtual Resort"*, they are getting some type of enticement to come back and spend more money within the municipality. Restaurants and bars capture more customers, events get better attendance and the municipality realizes more in local sales tax and other revenues. *eCOUPONS* steer customers from one venue to another. **(See CHART 15-1)**

This is the future of municipal economic sustainability, not higher cigarette taxes, liquor taxes, or exorbitant parking fees.

CARLINI-ISM: *"Those cities which keep horse-and-buggy taxes and penalties as one of their revenue streams, will become economic ghost towns."*

CHART 15-1 LAYERS OF AN IREC COMPLEX

CREATES THE "VIRTUAL RESORT" EFFECT			
CONSUMER DEMOGRAPHIC COLLECTIONS			
CROSS-MARKETING APPS –eCOUPONS, eDISCOUNTS			
BROADBAND CONNECTIVITY (WiFi – DAS)			
VENUE1	VENUE 2	VENUE 3	VENUE 4
BUSINESS ENTERPRISE ZONE			

WE DON'T NEED CAPABILITIES. YES, YOU DO.

In a blog, Mark Cuban, the owner of the Dallas Mavericks (NBA Team) and one of the "visionary" panelists on the TV reality show Shark Tank, discounted the need for adding wireless capacity to stadiums and instead said people should focus on the game and not on their Smartphone.

> *"As in every business you have to always ask yourself what your product is and the best way to deliver it. In the NBA our product is fun and energy. The last thing we need to do is encourage our customers to stare at their phones.*
> *I can't think of a bigger mistake then trying to integrate Smartphones just because you can …..The fan experience is about looking up, not looking down. If you let them look down, they might as well stay at home, the screen is always going to be better there."*

For someone who made all his money on technology, it is funny to observe how he is now discounting the application of technology as Smartphones work their way into everyday life and sporting events. You cannot discount the fact that a growing number of people are using Smartphones.

It would be much smarter from a marketing standpoint to understand the technology and its applications to see how the devices can benefit your business, than to discount them before understanding their power and growing ubiquitous-ness.

In a diametrically opposed core business strategy, the NFL commissioner, Goodell, says they (the NFL) will have all stadiums supported by this new capability. The NFL Commissioner gets it.

CARLINI-ISM: *"Leading-edge convention centers, stadiums, & entertainment venues will not maintain their position using trailing-edge technologies."*

COVERAGE VERSUS CAPACITY

There is a huge difference between a network engineer looking at providing an area with "coverage" versus "capacity" when it comes to a room, a conference center, or a section of a stadium. Coverage means you reach the space and provide access to network service, but capacity means you provide access AND capacity even when demand is high.

In an entertainment environment like a stadium or theatre, how much capacity should be available? Should it be a matter of taking the maximum capacity of the venue and providing some type of partial capacity for peak demand? At what capacity should the service be geared for? 50% of full capacity? 25% of full capacity? 10% of full capacity?

These are the questions designers are asking today as no true rules-of-thumb have been established. Many first iterations of "smart parks" or *Intelligent Retail Entertainment and Convention Campuses* are still feeling growing pains as they adjust for capacity and continued growth of the use of Smartphones. Remember Smartphones are becoming ubiquitous. Everyone will have one shortly and some applications which are video-based, are very demanding of bandwidth.

The whole idea of Master Planning the facilities must be augmented to include the layer of connectivity and power applied for the benefit of the building owner, the building tenants, and their clientele. Municipalities must understand this as well if they want to attract and maintain the right venues within their economic domain.

COORDINATED SOFTWARE for "SERVICES AT YOUR FINGER-TIPS"

In a multi-venue environment, connectivity between all the entertainment venues and restaurants should be configured.

A VIP-type of atmosphere developed through the use of wireless technology and applications (apps) focused on a richer consumer experience to alleviate all the venues and restaurants to a higher level of consumer participation must be implemented.

What are we going to provide as *"services at your fingertips"* to the entertainment/ retail consumers and the convention attendees of the venues? This is what needs to be decided, outlined, and implemented by anyone undertaking this type of endeavor.

From a software standpoint, apps need to be in place to take the requests from the users' Smartphones and connect all the venues together with the consumers. (**See CHART 15-2** – Framework of Multi-Venue Connectivity below)

CHART 15-2: FRAMEWORK OF MULTI-VENUE CONNECTIVITY

LEVEL	DESCRIPTION	RESPONSIBILITY
VENUES' SOFTWARE	Applications for reservations, tickets, purchasing merchandise, concessions, customer services, etc.	VENUE SOFTWARE DEVELOPERS
APPLICATION LAYER (APPS)	SOFTWARE THAT IS THE INTERFACE BETWEEN THE CONSUMER (Venue guest/ attendee)	APP DEVELOPERS
TRANSPORT LAYER	CARRIERS' NETWORKS	NETWORK CARRIERS
PHYSICAL LAYER	ANTENNAE, NETWORK EQUIPMENT	SYSTEM INTEGRATOR, MUNICIPALITY

Source: James Carlini

Both the network PHYSICAL and TRANSPORT layers must be in place as well as the apps in the APPLICATION layer which will handle the users' requests.

Some software is already available and can be part of the solution. Other software must be developed to give a unique mix of capabilities. These can be discussed at the planning stage and then be developed as time goes on.

There is "stadium software" already available and those existing types of apps should be reviewed to discover if any can be adapted for your own situation. If you are working with a stadium, most teams already have their own apps.

What needs to be reviewed is, "Where are the improvements in electronic (Smartphone) capabilities to further enhance the team's fan experience?"

"Intelligent amenities are more important than traditional amenities. Traditional amenities are a given in commercial real estate – intelligent amenities are not."

"A single connection equates to a potential single-point-of-failure. That being said, over 95% of buildings are then not ready for tenants who have mission critical applications which need network access."

Why? More people are using Smartphones and Tablets than PCs today. They don't want to be burdened with "computing."

That sounds too technical. All they want to do is make a call and get things done.

DISTILLATION OF APPLICATIONS TO APPS

Middleware, firewalls, hosted services, infrastructure, and all the other complex facets of data-centric applications need to be re-engineered and streamlined to fit into the new choice in edge technology. No, it's not for the next laptop or PC. It's for the Smartphones and Tablets.

People don't care about all the technical interfaces, the connectivity concepts, or any of the other techie jargon of making applications work. All they want to do is pull out their Smartphone and complete a transaction. No wait, that is even too techie. All they want to do is make a "Cloud call."

What is a Cloud call? A Cloud call can be anything. It can be:

- pulling up a video,
- paying for an item (digital wallet),
- doing a Google search,
- asking for directions,
- taking and sending a selfie,
- making reservations,
- playing a game, or
- just making a phone call.

Some social media "tools" might still be searching for relevance and a better way to monetization, but what they have done is bring in a very casual type of user who isn't enamored by glitzy tech issues or complex interfaces.

The casual user just wants to communicate with others and facilitate transactions to make life simpler. They don't want to earn a degree in computer science in order to use their Smartphone or play "Angry Birds."

Convergence Of Edge Technology

When you go out of the house, do you always take your laptop? Do you always take a tablet? No, the only device you constantly take out is your Smartphone.

If the Smartphone is going to be viewed as the new "universal" edge technology, then all applications should be designed with that in mind. This means adjusting to certain issues and weaknesses, like power consumption, bandwidth, and screen size to display information.

There are already big shifts in re-working real estate and intelligent buildings so Smartphones can be applied as edge technology where no interfacing end-user technology exists.

An example is a Sports stadium. Going to the game, no one is lugging a laptop or toting a tablet, but they are bringing in their Smartphone. If the Smartphone can be used as the "edge technology" for interactivity with the stadium's network, look at all the apps which can be developed for a richer fan experience.

This is happening already and you have the ability to order concessions, order team wear and other event memorabilia from your seat.

This is breathing new life into the stadium as well as the teams' revenues without having to build out a total network infrastructure to a 70,000-seat stadium. Could you imagine the cost of installing an electronic interactive terminal device at every seat and hooking it back up to a group of servers?

Next-generation real estate and commercial buildings must take all this into consideration as people adopt new "edge technology" which is not tied to or limited to the company they work at. *"Let your fingers do the walking"* has been replaced with, *"Let your fingers do the shopping."*

CARLINI-ISM: *"Simplicity for end-user use, requires complex infrastructure and software that no one really sees (or cares about)."*

As we become a more "mobile" society, more credit cards are being embedded with an RFID chip in them to make financial transactions easier to execute.

An RFID (Radio Frequency Identification) chip is not the same type of chip as the NFC (Near Field Communications) chip is on your Smartphone (like in a SAMSUNG III, IV, or V – Apple iPhones did not have an NFC chip in them until the iPhone 6). Smartphones may start to replace credit cards as the financial instrument in retail transactions.

NFC chips are used for "mobile wallet" applications using your Smartphone instead of a credit card. The NFC chip can be encrypted. The RFID chip on your credit card cannot be. Here are some differences between the two technologies. Below shows a table of comparisons and differences of RFID chips and the NFC chip: **(See CHART 15-2)**

CHART 15-2: RFID CHIPS vs. NFC CHIP

CHARACTERISTIC	RFID CHIP	NFC CHIP
Usage	In credit cards, asset tags, other inventory IDs for supply chain management, tool management, materials management, access control, attendee tracking (Conferences).	In some Smartphones. (mobile wallet) Also now out - NFC tags for new marketing apps.
Transmission	One-way only.	Can be two-way.
Signal	Always on (provides info any time it is scanned).	Must be activated.
Range	Several feet to 300 feet. (100 meters)	Only 10 cm. (four inches)
Encryption	Cannot be encrypted.	Can be encrypted.
STORAGE	2 Kilobytes – 128KB	64 bytes – 8KB
Scanning Capability	A scanner can read multiple chips at once.	Only one at a time.
SPEED	Effective scanning rate goes into 160+ MPH.	Up to 424kbps.
Frequency	LOW - 125-134 KHz HIGH – 13.56 MHz Ultra HIGH - 856 - 960 MHz, 3-10Ghz.	13.56 MHz
Capability	Can only be used as a Tag.	Used as Tag or Reader. Can communicate peer-to-peer.
OVERALL SECURITY	Less secure.	More secure.

Source: James Carlini

CHAPTER REVIEW

CARLINI-ISMS

"Let your fingers do the walking, has transformed to, 'Let your fingers do the buying,' with today's Smartphones."

"Those cities which keep horse-and-buggy taxes and penalties as one of their revenue streams, will become economic ghost towns."

"Leading-edge convention centers, stadiums, & entertainment venues will not maintain their position using trailing-edge technologies."

"Simplicity for end-user use, requires complex infrastructure and software that no one really sees (or cares about)."

[16]

INTELLIGENT INFRASTRUCTURE: THE FUTURE

"Everyone faces the future with their eyes firmly on the past and they don't see what's going to happen next." – **JOHN ROBINSON PIERCE, Bell Labs Scientist and a father of the transistor and satellite communications (From the book – The Idea Factory: Bell Labs and the Great Age of American Innovation.)**

*"B*uilding for the future means advancing from the past."* This implies looking at challenges differently and not using the same out-of-date, rules-of-thumb, last century methodologies, or other inapplicable metrics when trying to be a trail blazer into new territories (both physical and conceptual).

When approaching a new dynamic area in infrastructure, real estate, or technology, a whole new framework for analysis may be necessary. Too many people try to fit something new into a traditional framework or traditional analysis model which does not fit as a metric.

We are at a huge crossroads of both transformation and convergence.

Looking into what is needed for a solid platform for global commerce, the main issues are to offer the best and the fastest capabilities. Speed is always a factor in any transportation medium, including communications. Anyone, or any corporate or political sales pitch, which tells you otherwise is trying to protect obsolete infrastructure and poor competitive strategies.

Quality of the network infrastructure has to also be firmly embedded in the design because in "shipping and delivering" information (voice, data, and video), you cannot deliver the information all garbled up. The more time we waste in approving and implementing these new "super-delivery" network infrastructures, the more we fall behind in global competitiveness.

We have to move swiftly in upgrading our network infrastructure. And if you have read up to this point, you know: "Time is not money, time cannot be replaced."

In the book, **The Idea Factory**, the author John Gerstner addresses the problems which occur when trying to apply new technologies to business as well as society in general. He believes:

"..... inventions don't necessarily evolve into the innovations one might at first foresee. Humans all suffered from a terrible habit of shoving new ideas into old paradigms."

The point in his observation should be well-taken. Many times, instead of assessing the current framework and saying, *"Hey, we can just build onto it"*, the observation and strategic directive need to be:

"What do we need to modify, add on or totally get rid of, in order to incorporate this new capability into the current environment?"

When it comes to building a network to support Smartphones and those yet-to-be-developed devices following them which will support the *"Internet of Things"*, you may not able to engineer the new network to be backwards-compatible with the old network architecture. You may need to take all the old equipment out and install a whole new system. This approach is not the one used by incumbent phone companies. They like to make everything "backwards-compatible" which used to be the design principle in the 20th century but today, that approach might be a losing strategy.

Expensive? Yes, but what are the long-term impacts? Installing everything new, instead of supporting both old and new technologies in a parallel environment, may turn out to be much cheaper.

CARLINI-ISM: *"The Internet of Things needs to be able to run on the Internet of Reality (the network infrastructure that is in place)."*

REAL GREEN INITIATIVES

I was told a new type of RJ45 connector was being designed at the time of writing this book and the modifications to it would run at 75% less power consumption (one watt per connector instead of four watts). It becomes a significant factor in data centers and areas where there is a large concentration of these connector-types.

The example given to me was one of the large government data centers. The Social Security Administration has a data center with 500,000 ports in it. If you can reduce the power consumption of all those connectors by 75% just by switching out the RJ45 jack to its next-generation, look at all the power you save just with that one change.

500,000 times three watts is 1,500,000 watts of power (or 1.5 Mega-watts of power) you have just reduced one data center by.

That is the type of real energy savings facilities managers and data center managers have to sit up and take notice of. Too many "green energy" advocates are trying to sell property owners and facilities managers on sophisticated lighting systems with special dimming sensors that have twenty-five year paybacks. They don't understand the market. No one is going to buy off on that type of savings package or that long a pay-back period.

On the other hand, with so many companies having data centers and mission critical call centers, switching out such a basic component as an RJ45 jack will give a 75% reduction in power consumption. That is what real cost savings is all about and what is going to spark interest in the people who manage properties.

Too many people are running around with ideas of what networks will be like in ten or twenty years. Their focus is not built on any long-term work on network infrastructures, but more on press releases and superficial social media announcements focusing on off-the-wall applications which will never become popular or mainstream. What they need to understand is, nothing will be adopted if it cannot run on the network infrastructure that is in place. As I said earlier in the book, *"The Internet of Things can only run on the 'Internet of Reality.' The actual network infrastructure that is in place."*

When I was at Bell Telephone Laboratories, I was able to work with many talented people. I remember one senior engineer saying something prolific when we were discussing how to re-work an awkward software package for users to work with. His statement got etched in my mind and I have referred to his observation many times when talking with people about designing new systems, hardware and user interfaces. The engineer, George Schumacher, said, *"The technology should adjust to us, we shouldn't have to adjust to the technology."*

His observation still holds true today as we see many people trying to "adjust" to make systems work right within their organization when the system itself should be re-worked so it is compatible with the people using it.

CARLINI-ISM: *"Good technology should adjust to us. If we have to adjust to use the technology, it is no good."*

WHERE DO THE IDEAS COME FROM?

Doing something different or adding something new may not come from those who are immersed in current endeavors. New ideas may come from those outside of the industry because they are looking at things from a different perspective and have not yet adopted (or been saddled with) the traditional framework that is "accepted" by the industry and its "experts."

Sometimes adopting the traditional framework of the industry, locks out all future creativity and innovation from the outside as well as the further explorations for a different approach.

A long time ago, an executive from 3M told me about new product development, "Evolution comes from the inside. Revolution comes from the outside." He gave the example of cars and transportation.

The tire companies all used rubber as part of their products so they were always trying to refine the tire as well as its use of rubber in its application.

Through several decades of use on the automobile, the product evolves from rubber tires with inner tubes to tubeless tires to radial-belted ones with steel belts in them for durability.

He pointed out, that is an evolution of a product all from within the industry. Outside the industry, no one is looking at perpetuating the use of rubber in tires, they might ask why doesn't the vehicle ride on a cushion of air and get rid of the tires altogether?

All the tire companies, who have vested interests in rubber plantations and related manufacturing processes, are not going to get rid of the tire. They are just going to refine it and make improvements on it. Those involved in the process may not even see there is any alternative approach and they definitely do not see any reason to even ask the question about any alternatives being viable.

CARLINI-ISM:

"Operational people are not futurists. Do not listen to those working in day-to-day operations because those who work in the trenches do not see the horizon."

SEEING THE BIG PICTURE

There was a poll taken on one of the social media tools (LinkedIn) while this book was being written which asked respondents what they thought was more important as a technical challenge for the next several years – Big Data or Real Time Analytics?

You could choose one or the other, both or neither. I chose "neither" because after reading the comments by the respondents, they seemed too focused on their own little niche of expertise, instead of focusing on the big picture. (A common problem)

Before you do anything with Big Data and Real-Time Analytics, you need to make sure the infrastructure you are using can handle and transport all of this information. It doesn't. First thing is to make sure the network infrastructure is in place (totally) and then you can add any application you want. Anywhere.

Today, the network infrastructure is not consistent. One Gigabit per second (1Gbps) should be the baseline speed down to the end-user on wire or wireless. With Smartphones and Tablets becoming the ubiquitous "edge technology" people seem to be quickly adopting, we need to have a network infrastructure which can deliver huge amounts of data (or video) down to that end-user.

With more video-based applications becoming popular, multi-gigabit speeds to the end-user will be necessary if you are talking about Big Data as well as Real-Time Analytics. So if we are "prioritizing" technology challenges which are obstacles to us moving forward, we need to take a much broader view of things that are critical.

Again, that is one of the bigger challenges: Getting people to develop and apply a much broader perspective than the narrow one they graduated with in their Computer Science, Engineering, or "fill-in-the-blank" degrees.

CARLINI-ISM: *"Tomorrow is not the time to start building for tomorrow. Future viability tomorrow depends on building a solid infrastructure today."*

Looking Forward – Accelerating The Speed Of Business

We need to upgrade the entire network infrastructure layer within the global *Platform for Commerce*. It is a major task and not an easy one to sell to those executives and government leaders who still don't understand how all the layers within the infrastructure are critical to regional economic development.

What is the "speed of business?" If we were living in the times of Columbus, the fastest way to get something from one place or the other was by sea. It could take weeks or months to get a letter or a package from one point on the earth to another.

As time progressed, so did the speed of business. When the telephone came into the economic landscape, the speed of business was measured by how long railroads took to bring mail from the East Coast to the West Coast. This was a huge improvement over Columbus's day. As the telephone network grew, the speed of business accelerated.

Now, with the Internet and ultra-high-speed servers as well as end-users moving up to Smartphones, the need for higher speeds to the end-users is critical. The "speed of business" can be almost instantaneous.

Stock market traders already use high-powered HFT servers which can place trades every several microseconds. They understand shaving off a couple of microseconds can give a great competitive advantage in trading securities. They are reaping the economic benefits on every trade they make from the super speeds their servers are running at.

We need to uncover and exploit new ways of accelerating the business process in other areas. When we do, we will realize a benefit and a real explosion of new applications which before were technically infeasible or just viewed as "impossible."

Think about the collection and distribution of criminal fingerprints. Each set of fingerprints takes up about six to ten megabytes of storage. If you have 300,000 to 350,000 sets of prints, it is a significant amount of storage.

If you want to share a big data base, you need to have very fast speeds in the network in order to deliver that content. This would have been an impossible application to implement years ago, but with today's high-speed networks, it is a no-brainer.

Increasing network speeds open up a whole new class of applications across all industries. Pick any industry, and you will see how higher-speed network infrastructure brings in new and innovative ideas which expand and extends the range of practical applications. **(See CHART 16-1)**

An article I wrote on Terabit speeds got mixed reviews. Some people thought this was ridiculous to be touting terabit speeds when most people are still looking at multi-megabit and single gigabit solutions. I can't help it if the velocity of technology change and smart device adoption is accelerating, and the need for much faster broadband connectivity is creating a bigger gap between installed networks and planned networks. **(See CHART 16-2)** Bottom line, we need to get things moving faster.

CHART 16-1: NEW APPLICATIONS BROUGHT IN BY FASTER NETWORK SPEEDS

Law enforcement	Distribution of sets of finger-prints, remote video surveillance of high-crime and high-value targets. More real-time surveillance.
Medical Industry	High-speed transfer of X-rays and MRIs. High-speed transfer of medical images. Remote surgeries.
Manufacturing Industry	High-speed transfer of complex diagrams and AUTOCAD. 3-D Manufacturing. Real-time quality control management using sensors across the whole process.
Communications Industry	Faster switching technologies, higher bandwidths down to the end-user (streaming video).
Financial Industry	HFT (High Frequency Transac-tions) Faster trading rates.
Military	Target Acquisition, real-time battlefield coordination.

It was interesting as there was some negative reaction by those who thought it was too big a leap, but it is exactly the reason to push forth this idea of raising the bar to Terabit speeds for those networks in the planning process.

We need to build a network infrastructure which can last for more than two or three years. The speed of adoption of new technology is accelerating and that adoption is also creating the need for much faster networks which can handle a bigger capacity of traffic.

CHART 16-2: CLARIFYING SPEEDS

SPEEDS	MATH EQUIVALENT	SOME FACTS
KILOBIT	**Thousands** (of bits per second)	In 1981, 9600bps was considered a "fast network." A 9600bps Modem cost around $7,000 back then.
MEGABIT	**Millions** (of bits per second)	The first T-1 (1.544Mbps) was installed in 1963 in Skokie, Illinois. Corporate access and availability came much later. A T-1 is a half-century old technology.
GIGABIT	**Billions** (of bits per second)	Some said this was crazy several years ago. Now, we already have 40Gbps going into Business Parks.
TERABIT	**Trillions** (of bits per second)	Need to keep sights set on emerging technology that will make Terabit speeds obsolete.

We are adopting and accelerating speeds faster than what some people think and to design a network to handle MORE traffic is the right thing to do. It would add to the length of time the network would be viable to handle growing traffic loads.

TERABIT SPEEDS? NOW IS THE TIME TO RAISE THE BAR

We are moving forward at a fast pace. Terabit speeds will be adopted for network backbones faster than most realize. There are already tests on fiber to run multi-terabit speeds.

We are already beyond the "testing in the lab" stages. There are companies like Alcatel-Lucent and British Telecom who have already tested terabit speeds in real environments. They tested a 255-mile run of fiber to see if they could attain Terabit speeds and they did. In the lab, they attained over 30Tbps (Terabits per second) and in the street they already have passed 1.4Tbps.

How else are networks going to able to keep up with the deluge of devices being predicted for the year 2020? They have to increase speeds by a geometric amount.

If you are planning a network as you are reading this book, you should be setting your goal to terabit speeds, not multi-gigabit speeds. (At least, for the backbone network.)

There is a new concept being developed called **5G Networks**. It is the next big step in network infrastructure moving everyone from 4G networks to 5G networks with the target date of 2020 for deployment. Major enhancements found in 5G Networks are listed on **CHART 16-4**. Let's examine this table and clarify the numbers.

When the Bell System was in charge of laying down copper cables to service customers in an area (like a downtown area), their design plans were to install enough cable to last 20-30 years because the big part of the cost was the excavation of the trenches to put the cable in. They did not want to go in and dig up cable routes every couple of years. In those days, the velocity of more subscribers using new devices and video applications were not driving the growth in traffic.

CHART 16-4: 5G NETWORKS – MAJOR UPGRADES & ENHANCEMENTS

FEATURE	DESCRIPTION
HIGHER DATA RATE	Projected 10-100 times data rate increase. Using mix of high & low frequency bands. Spectrum efficiency is essential.
HIGHER SYSTEM CAPACITY	Focused on being 1,000 times more capacity.
ENERGY SAVINGS/ COST REDUCTION	Less power used. New antennae configurations. (LRC – Light Radio Cube)
MASSIVE DEVICE CONNECTIVITY	100 times the amount of connected devices.
REDUCED LATENCY	Aiming for less than 1ms. (one millisecond)

It was pretty much a network focused on carrying voice traffic that grew at a relatively slow rate. Those wanting us to take down to a slower development process are either:

1) Trying to protect obsolete infrastructure and obsolete design ideas (typical of the incumbent phone companies), or

2) Unaware from a global perspective, we need to become more of the "leader of the pack" rather than playing catch-up to the pack, or

3) Brainwashed from the propaganda generated by those in number 1, the incumbent phone companies.

So ask yourselves, where are you when it comes to understanding the needs for a state-of-the-art network infrastructure which can support your local economic development goals?

Do you have any detailed plan in place? If none, you are behind those who think establishing a goal of "gigabit speeds" is satisfactory. It isn't.

Any network being planned today for tomorrow should be conceptualized like this. **(See CHART 16-5)** Let's examine this table and clarify the numbers.

1) These suggested speeds are for networks which have yet to be installed. If you are going to build something, at least build something that should last for awhile.

CHART 16-5: DESIGN CRITERIA FOR NETWORK SPEEDS

TYPE OF USE	SPEED	DESCRIPTION
Common End-User/ Subscriber	**1Gbps (One Gigabit per second)**	This includes wireless due to what Smartphones are demanding in bandwidth.
Industrial Park, Business Campus Commercial Space	**40-100Gbps**	This would include next-generation Intelligent Business Campuses. (Some parks already have multiple carriers providing 40Gbps today.)
Downtown/ Commercial Space	**40-100Gbps**	For downtown urban areas.

2) For next-generation Intelligent Business Campuses/ Intelligent Industrial Parks what you offer at any one location is going to dictate what gets puts in (i.e. when you don't offer high enough access speeds, certain corporate site selection committees will pass you by depending on what they are looking for.) So if you do set your sights low when it comes to speed, you won't be able to land corporate facilities you think will move into the business park.

3) As Smartphones and Tablets become more ubiquitous, demand for speed (for new apps) will increase and not go down. Some new installation endeavors at venues, like stadiums and ball parks, have already shown they are under-engineered. Current engineering "rules-of-thumb" do not reflect actual demand and need to be re-thought. If anything, they (those in long-term network engineering - including those at the carriers) have to leapfrog what is already a current market condition.

CARLINI-ISM: *"Speed opens up new applications. What was impossible, now becomes possible."*

A QUICK LOOK AT 5G NETWORKS

5G Networks are being discussed and better defined by key industry manufacturers and network carriers so they can be deployed in the 2020 time period. They will represent a giant leap in capacity, baseline speeds and overall capabilities. They have to. Consumers are turning to Smartphones and Tablets as the new ubiquitous edge technology for all their applications.

Many people have gotten a good taste of Smartphones and all the applications they can utilize with them. Now, the growing demand is established and both the framework of networks as well as their capabilities must be upgraded dramatically to keep up with user demand and exponentially growing traffic.

When approaching a new dynamic area in infrastructure, real estate, or technology, a whole new framework for analysis may be necessary. Too many people try to fit something new into a traditional framework or traditional analysis model which does not fit as a metric.

Looking into what is needed for a solid platform for global commerce, the main issues are to offer the best and the fastest capabilities. Speed is always a factor in any transportation medium, including communications. Anyone, or any corporate or political sales pitch, who tells you otherwise is trying to protect obsolete infrastructure and/or poor competitive strategies.

There needs to be good educations of all interested parties, instead of having some people hypothesize concepts which will never be prevalent or even applicable. Remember, the network infrastructure is part of the **Platform for Commerce**. The **Platform for Commerce's** five millennia focus has always been on increasing trade routes and commerce.

BANDWIDTH

Any network being planned today for tomorrow should be designed to handle appropriate speeds for a variety of users. **(See CHART 16-6)**

5G Networks will be providing some very high speeds to the average user. This means building out the network to specifications which encompass terabit backbones.

Those network architects in long-term network engineering, including those at the incumbent carriers, must leapfrog what is already an under-engineered configuration target.

CHART 16-6: DATA RATE FOR 5G NETWORKS

SPEED	MARKET SEGMENT
100Gbps +	Specialized enterprise users (stationary)
50Gbps	Low-mobility users
5Gbps	High-mobility users
1Gbps	Anywhere (baseline speed)

Having more bandwidth available will accelerate the amount of applications which are feasible for customer service, video, high-definition video, social networking, and so many other applications.

In addition to being aware about the growing demand for speed, become more aware of new building blocks for technology as well as the infrastructure and real estate. The faster we can apply cutting-edge solutions to problems, the more competitive we become in the global markets.

CARLINI-ISM: *"The future belongs to the swift and the decisive."*

This gives you a clearer picture as to where network infrastructure is evolving. **(See CHART 16-7)**

CHART 16-7: 5G NETWORK ELEMENTS

COVERAGE	This was an initial concern of network designers for 3G and 4G.
CAPACITY	This became a bigger concern as 4G networks got tested by users' applications and increased traffic.
CONTINUUM OF SERVICES	How long will services last before being replaced or discarded?
CAPABILITIES	This is constantly evolving and will impact the above elements. Influencers in this level include manufacturers, software/ app designers, and customer demands.
COMMUNITY COMMITMENT (COST)	Think of this as the political, regulatory, and regional economic development link for the network. User demand will shape the network, but so will strong regulatory guidelines that could hamper or promote network resiliency and capabilities. Smart communities will realize that having a solid network infrastructure will directly translate into having a solid economic base.

CARRIER COMMITMENT (COST)	This is the carrier's focus as to what it wants to provide to the region: The most advanced, the status quo, or trailing-edge capabilities because they don't see a market or a big return on their investment.
BACKHAUL (BACKBONE)	This is the foundation that must be strong and resilient enough to handle everything put on top of it.

Source: James Carlini, All Rights Reserved

We need to focus on getting back in front of the pack when it comes to network infrastructure and moving to Terabit speeds can only open up a whole new class of applications which can only be feasible when networks run at terabit speeds. For the future, we need to strengthen the resiliency of networks as well as increasing their speeds.

GRAPHENE – THE NEXT BUILDING BLOCK LIKE THE TRANSISTOR?

The weaknesses of the technologies running in Smartphones are also being addressed and strengthened as this book gets published. New "building-block" materials, such as, graphene are being researched, developed and tested to be used as the foundation materials for next-generation photo-detectors, batteries, micro super-capacitors, and many new innovations in Smartphones.

Graphene is a one-atom thick, mesh-like (think hexagonal honeycomb or chicken wire), semi metal that will add more battery-life into a Smartphone among other breakthrough ideas like bendable displays.

It is also ten times stronger than diamonds so its resiliency is perfect for use in a Smartphone or any other device that requires super-ruggedness.

Wikipedia defines it as:

High-quality graphene is strong, light, nearly transparent and an excellent conductor of heat and electricity. Its interactions with other materials and with light and its inherently two-dimensional nature produce unique properties, such as the bi-polar transistor effect, ballistic transport of charges and large quantum oscillations.

Is this a new miracle micro-metal or film that will become a universal coating for many devices? It will definitely become a building block for next-generation electronic and communication devices as well as quick-charge batteries. IBM has already tested it in a computer chip and it runs 10,000 times faster than previous chips with graphene added to them. In the past, graphene has been hard to mass-produce, but that obstacle is being worked on by Samsung.

This will definitely affect the development of Smartphones and give new capabilities as well as extending what is now considered their big limitations, like battery life and rigid phone frameworks. Samsung is doing a lot of research with this new capability. If they can mass-produce it cheaply as a building block for their Smartphones, they will definitely raise the ante in the big war of innovation with Apple.

CARLINI-ISM: *"Good competition accelerates good innovation whereas monopolies stifle innovation and protects against change."*

In an everyday example, how long does it take for you to re-charge your Smartphone? Several hours every couple of days? What if you could do it in 30-40 seconds and then not have to re-charge for several weeks? That is the difference graphene can make just in battery life. Just think if one phone maker gets that capability and the others can't.

Research has already demonstrated that graphene, which is made up of carbon atoms, is capable of absorbing 90 percent of electromagnetic energy across a high bandwidth. Also, it is like putting on a micro-Kevlar material to protect against EMI (Electro-Magnetic Interference). This could be used in many ways.

Think of graphene being used as a one-atom thick, micro-privacy drape that could filter out EMI from car windows or buildings to eliminate EMI interference. This could be a new building block for secure wireless networks and other applications needing EMI filtering.

Using this type of EMI "graphene drape" could be a good way to secure trading floor networks in all the stock exchanges as well as other mission critical wireless networks found in so many industries today. Because graphene is transparent, offices could have this applied to windows so no one could intercept signals being emitted by a wireless router being picked up by a directional antenna.

CARLINI-ISM: *"Is graphene the silver lining for cloud computing? When it is perfected, it will be as common as a transistor in many devices, edge technology, and high-tech real estate."*

This new material could definitely be viewed as an "Intelligent Amenity" for 21st real estate, especially in those buildings wanting to offer security from electronic eavesdropping. New buildings would have it as part of the design spec and existing buildings could add it if they wanted to compete for security-conscious corporate tenants.

Graphene can be used in many applications including solar cells. Its capability to conduct heat and electricity makes it a quality building-block material for many devices and solutions. In a couple of years, it will be looked at as an integral component for a lot of products across many industries.

Besides Samsung and IBM, Johnson Controls, Tesla, and Lockheed-Martin are all researching how they can use this material in the products they make as well as the products they are developing.

To prepare for the network infrastructure of the future, be cognizant of the speed of adoption of new devices by the average consumer and new bandwidth-hungry applications which are putting huge demands on current networks. Set the design bar high when it comes to network speeds.

In addition to being aware about the growing demand for speed, become more aware of new building blocks for the technology as well as the infrastructure and real estate materials. The faster we can apply cutting-edge solutions to problems, the more competitive we become in the global markets.

CARLINI-ISM: *"What is considered 'the ultimate' today sometimes becomes obsolete tomorrow - literally."*

LEADING/ LEAVING WITH A PRAGMATIC PERSPECTIVE

Moving forward into the future and determining how all these paradigm shifts affect you and your job, remember to be pragmatic. Do not buy off into all the shrill sales pitches being hurled about in the market. You are much better armed with a multi-faceted skill set than you had before. **(See CHART 16-8)**

Just like in courses I taught and stated in the beginning of this book, I want to leave you with this: ***There are no experts in this industry, the best you can be is a good student – always learning.***

If you learned anything from this book, you learned applying technology is a continual process and not a "one-shot deal." Learn from others, both their successes and their failures and you will be rich in skills and wisdom.

And here is something I learned at MIT while taking courses there for AT&T: Don't learn and memorize facts. Learn how to learn. Learn where to go to get new information on the next big thing.

CARLINI-ISM: *"If you learn from other cultures, you can add all their wisdom and knowledge to your set of skills. That is real power."*

CHART 16-8: SKILLS YOU SHOULD BE ARMED WITH BY NOW

TOOL	SKILLS GAINED	APPLICATIONS	FOUND IN CHAPTER
CARLINI-ISMs	*WISDOM, LOGIC, OUT-SIDE (second) PERSPECTIVE*	Everyday Life.	End of every chapter or 17
PLATFORM for COMMERCE	*Getting a full, clear picture of what infrastructure is and how we got to this point.*	Everyday life. Understanding the impact of infrastructure on real estate & regional economic development.	5
Five-Level TARGET MAP of TECHNOLOGY	*Structured review of technology within the enterprise and organization.*	Review of budgets, in-place technology and how it all relates to the business.	9
RFP GUIDELINES	*Smarter Buying & Selling of Technology. More structured approach to buying technology & getting away from Price as a single criteria.*	Comparing and buying equipment and services for day-to-day as well as long-range strategic applications.	10
FOUR LEVELS OF SELLING	*Understanding where skills are needed, applied and paid for.*	Buying & selling of high-tech equipment & services. Comparing others.	11
LEADERSHIP	*Better understanding of what leadership is and should do.*	Management of people and resources. Making critical decisions.	12
UNDERSTANDING NEXT GENERATION REAL ESTATE	*Understanding the differences in Master Planning and requiring Intelligent Infrastructure*	IBCs, IRECs, upgrading older business and industrial parks.	8, 14, 15

CHAPTER REVIEW

CARLINI-ISMS

"The Internet of Things needs to be able to run on the Internet of Reality (the network infrastructure that is in place)."

"Good technology should adjust to us. If we have to adjust to the technology, it is no good."

"Operational people are not futurists. Do not listen to those working in day-to-day operations. Those who work in the trenches do not see the horizon."

"Tomorrow is not the time to start building for tomorrow. Future viability tomorrow depends on building a solid infrastructure today."

"Speed opens up new applications. What was impossible now becomes possible."

"The future belongs to the swift and the decisive."

"Good competition accelerates good innovation whereas monopolies stifle innovation and protects against change."

"Is graphene the silver lining for cloud computing? When it is perfected, it will be as common as a transistor in many devices, edge technology, and high-tech real estate."

"What is considered 'the ultimate' today sometimes becomes obsolete tomorrow – literally."

"If you learn from other cultures, you can add all their wisdom and knowledge to your set of skills. That is real power."

[17]

CARLINI-ISMS

A SUMMARY OF ALL THE CARLINI-ISMS WITHIN THIS BOOK

CHAPTER TWO

"Whoever has the best-trained workforce, is the toughest competitor."

"The broader the perspective, the better the problem-solver, especially in applying technology to business."

"Eliminate any single-point-of-failure on any mission critical network and/or application."

"Always try to explain technical terms in everyday terms and you will get a lot more positive response as well as clear understanding from those who do not have a technology background."

"Structuring the unstructured is a large step in the planning and design of any mission critical application."

"Quality is the bedrock of all successful endeavors."

"Don't be politically correct. Be politically accurate."

"Courses aren't boring, people are. A good teacher can generate excitement about any topic."

"The use of traditional business advisors - lawyers and accountants - is not enough for assessing deals involving technology. They must be augmented with technologists."

"Intelligent amenities are more important than traditional amenities. Traditional amenities are a given in commercial real estate – intelligent amenities are not."

"A single connection equates to a potential single-point-of-failure. That being said, over 95% of buildings are then not ready for tenants who have mission critical applications which need network access."

CHAPTER THREE

"Creating new conceptual frameworks for infrastructure means tearing down obsolete education and curricula."

"Building for the future means advancing from the past."

" 'Those who can, do. Those who can't? Teach.' This has to be replaced with – 'Those who can – MUST teach, in order to promote a pragmatic, realistic perspective within the upcoming workforce.'"

"Leading-edge countries do not maintain their position using trailing-edge infrastructures."

"Organizational structures become obsolete just like the technology that they manage. Management structures must be reviewed and replaced just like systems, software and technology."

"New skills are needed to solve information age endeavors. Do not rely on the "traditional methods" of management."

"Today, many people do not know how we got to this point in applying technology. You have to know what led up to today's configuration and applications of technologies."

"Fads fade fast. Learn how to distinguish between toys and tools."

"Professionalism is not a degree or certifications; it is a state of mind."

CHAPTER FOUR

"Economic development equals broadband connectivity and broadband connectivity equals jobs."

"Municipalities are holding on to some white elephants, they just don't know it – yet."

"If you are looking for a building or park to support your enterprise's Mission critical applications, over 95% of buildings are technologically obsolete. They have single vendor connectivity which in turn creates a single-point-of-failure."

"If it takes a village to raise a child, then it takes a region to raise economic viability for all."

"There is no such thing as a universal solution. What works down the street may not work for your environment."

"Think of technology assimilation as being RARE. The four stages are Research, Acceptance, Regulation, and Enforcement of the regulation."

CHAPTER FIVE

" 'Building for the future'" is a very hollow statement, when funding is aimed towards maintaining the past."

"To better see into a complex future, a clear framework is needed to structure new concepts."

"1950s solutions will not solve 21st century economic problems."

"Good government fosters good commerce. Bad government fosters no Commerce and attracts crime."

"Best practices are a moving target. Best practices need to take into account applying technology. And, best practices change with the weather."

"He who controls the press, controls the rest."

"Best practices are not found in bureaucracies."

CHAPTER SIX

"Competition is good. Monopolies are bad. Having competitors accelerates innovation within an industry."

"1950s solutions will not solve 21st century economic problems."

"Best Practices are a moving target. And, best practices change with the weather."

"Network infrastructures are like icebergs. You only see 5% of the issues and responsibilities on the surface."

"Bandwidth is like garage space. The more bandwidth you have, the more you will fill up."

"The first T1 circuit was installed in Skokie, Illinois in 1963. Hardly, cutting edge technology today...."

"Don't say you're 'state-of-the-art' if you are still using horse-and-buggy rules-of-thumb for your building(s)."

CHAPTER SEVEN

"20th Century solutions will not solve 21st Century problems."

"A Smartphone application isn't any good if you cannot access it."

"In today's commercial building markets, high-speed network infrastructure within a building is an intelligent amenity that is a must have, and not a hoped for."

"You must understand the problem in order to apply the right solution."

"We are past the Industrial Age, past the Information Age and into the mobile Internet Age. Let's not reinvest in the past, when there is so much to do for the future."

CHAPTER EIGHT

"20th century real estate strategies and solutions will not fit or satisfy 21st century real estate requirements."

"What is put in first, dictates what is put in next when leasing up next-generation Intelligent Business Centers (IBCs). You should adhere to a theme."

"When states do not address the right issues, job erosion occurs and they lose their tax base."

"Municipalities which create a lot of red-tape and development fees for developers build more walls and obstacles. These obstacles kill economic development instead of streamlining processes which open up doors and invite local economic development."

"Mission critical applications require redundant power sources as well as redundant network services. Redundancy is NOT an option."

"Master planning today is not only a new ball game. It's a new sport."

"Any intelligent amenity eliminating 90% of the competition should be seriously considered a top priority for any real estate developer or owner."

"You must develop metrics and comparisons for high-tech real estate. If you don't, you will be measured by your competitor's yardstick and you will always come up short."

CHAPTER NINE

"You must understand the usefulness of the technology infrastructure servicing your enterprise."

"Applying network based information systems to sustain a competitive lead is not a one shot deal. It's a continuous process."

"Setting the standard is the sign of an industry leader. Playing catch up or succumbing to providing mediocrity is the sign of a member of the trailing pack."

"If an organization is stagnant, all it wants is someone to maintain the current systems – in effect, a network janitor."

"A well-trained workforce is a good hedge against the competition."

"Creative people are scarce and do not come cheap. When you pay peanuts, you get monkeys."

"Taking something which has been proven in one industry and applying it to another industry can accelerate the solutions to both simple and complex problems."

CHAPTER TEN

"There is no such thing as a new $5,000 Rolls-Royce. You get what you pay for."

"Conversely, if you only have $5,000 to spend, there is no such thing as a Formula One Yugo."

"You don't have to be an expert. You just have to ask a couple of detailed questions so the vendor thinks you ARE an expert and believes he/she better not try to sell you a bill of goods."

"In comparing options, you always want to have an objective yardstick."

"No system or technology runs at 100% reliability. Even the phone company has down-time within its network."

"If you cannot determine what downtime costs are on a per-call or per hour basis, you are not managing your call center or any network effectively."

"No vendor is going to tell you their system "is a bitch" to learn. They are all user-friendly."

"No system runs by itself. You need a systems manager or administrator."

"Beware of hidden costs and the penalty costs like the "Cost of Disruption" and the "Extra Cost to Upgrade" when projects have to be prematurely rebuilt because you did not think long-term."

CHAPTER ELEVEN

"Be the best, or get passed up by the rest."

"Technology should be viewed as an investment, not as an expense."

"A good salesperson knows their product. A better salesperson knows their competitors' products as well."

"Executive skills include knowing how to speak well and knowing how to write well."

"When the vendor's technical knowledge and support is shallow, don't bet your company – or your job – on them."

"You have to believe in what you are selling. If you don't believe in it, why should the customer?"

"Know when to walk away from bad business."

"Real estate firms are in for a big surprise. The buildings that lack solid intelligent amenities will lose first-class tenants and slip into second-class properties. It's happening already."

"Real estate, infrastructure, and technology concepts are converging. They should be presented in a comprehensive manner in order to instill a broader perspective within students as well as professionals."

CHAPTER TWELVE

"You manage resources, but you lead people."

"A leader can be a manager, but a manager may not be a leader."

"If an organization fails to recognize talent, it starts to lose it to the competition."

"Only performance is reality. Everything else is just glossy rhetoric and slick slogans."

"Empowerment? Spread the responsibilities, but then, spread the recognition and the rewards as well."

"You don't get superman by offering Jimmy Olson wages. Serious people get serious money."

"It's time to be politically accurate and NOT politically correct."

"Take action. Let someone else be patient. Waiting around does not get you anywhere."

"Don't shoot a challenging salvo across my bow and then cry when I fire back a 44-gun broadside sinking your little ship."

"Corporate facilities have to support the corporate goals and objective.

CHAPTER THIRTEEN

"If the infrastructure is built right, a region can attract and maintain new businesses. Businesses add to the tax base and tax rates will actually go down, making it an even more attractive area to locate a business."

"'Quaint 'Tourist towns' better have the intelligent amenities to support affluent tourists, otherwise the tourists will flock to another destination that is more 'with it' for supporting their technology needs."

"Prioritizing projects and getting ideas of who they will impact is an analysis worth spending some time on!"

"Any government agency impeding the innovation and solution to infrastructure problems (which are impacting regional economic development) should be held liable for infrastructure negligence."

"You cannot get the right answers if the right questions are not being asked."

"Prioritizing projects and showing the methodology used to compare them provides a clear answer to whether or not money was well-spent."

"If a new technology is proven in a major application, it is hard not to try it somewhere else. The risk is minimized."

CHAPTER FOURTEEN

"Where is everything going? Time is not money – time cannot be replaced!"

"Rule for government agencies: If we are all in the same boat, let's make sure we all have an oar in the water and are pulling together."

"Avoid offers from any incumbent phone company to 'provide the business park's network infrastructure.' If you do, you just choked off connectivity."

"New terminology and definitions need to be learned if you want to be able to design, build, sell, and operate real estate in the 21st century."

"Return calls. Nothing is more unprofessional than people who do not return calls. The call you don't return, is the call that could have made you $1,000,000."

CHAPTER FIFTEEN

"Let your fingers do the walking, has transformed to, 'Let your fingers do the buying,' with today's Smartphones."

"Those cities which keep horse-and-buggy taxes and penalties as one of their revenue streams, will become economic ghost towns."

"Leading-edge convention centers, stadiums, & entertainment venues will not maintain their position using trailing-edge technologies."

"Simplicity for end-user use, requires complex infrastructure and software that no one really sees (or cares about)."

CHAPTER SIXTEEN

"The Internet of Things needs to be able to run on the Internet of Reality (the network infrastructure that is in place)."

"Good technology should adjust to us. If we have to adjust to the technology, it is no good."

"Operational people are not futurists. Do not listen to those working in day-to-day operations. Those who work in the trenches do not see the horizon."

"Tomorrow is not the time to start building for tomorrow. Future viability tomorrow depends on building a solid infrastructure today."

"Speed opens up new applications. What was impossible now becomes possible."

"The future belongs to the swift and the decisive."

"Good competition accelerates good innovation whereas monopolies stifle innovation and protects against change."

"Is graphene the silver lining for cloud computing? When it is perfected, it will be as common as a transistor in many devices, edge technology, and high-tech real estate."

"What is considered 'the ultimate' today sometimes becomes obsolete tomorrow – literally."

"If you learn from other cultures, you can add all their wisdom and knowledge to your set of skills. That is real power."

GLOSSARY OF TERMS

"Abbreviations"

ACA –   *Affordable Care Act*

AHSC –   *American Hospital Supply Corporation*

ASME –   *American Society of Mechanical Engineers*

AT&T –   *American Telephone & Telegraph Corporation*

AY-   *Arthur Young*

BDD –   *Business Development District (similar to TIF District)*

BEZ -   *Business Enterprise Zone*

BYOD –   *Bring Your Own Device*

CAN-DO –   *Critical, Always Necessary, Desirable, Optional – Prioritizing Approach*

CEO –   *Chief Executive Officer*

CIO –   *Chief Information Officer*

CO –   *Central Office*

COD –   *Cost of Disruption*

CPE –   *Customer Premise Equipment*

CRE –   *Commercial Real Estate*

CTO –   *Chief Technology Officer*

DART –   *Designated Active Resource with Timeframes*

DAS –   *Distributed Antennae Systems*

DBC -   *DuPage Business Center*

DCF –	*Discounted Cash Flow*
DNTP –	*DuPage National Technology Park (now the DuPage Business Center)*
DOS –	*Disk Operating System*
DPP –	*Discounted Payback Period*
DSL –	*Digital Subscriber Loop (or Line)*
EMI –	*Electro-Magnetic Interference*
EMS –	*Energy Management Systems*
ESS –	*Electronic Switching Systems*
FACT or	
FACT-based –	*Flexibility, Adaptability, Creativity, & Technology Skills*
"Five Nines" –	*99.999 reliability*
FTZ –	*Free Trade Zone*
HC –	*Hidden Costs*
HFT –	*High Frequency Transactions*
~~**HR**~~	*Human Resources*
HVAC –	*Heating, Ventilation, And Cooling*
IA –	*Intelligent Amenities*
IBI –	*Intelligent Buildings Institute*
IBM –	*International Business Machines Corporation*

IBC –	*Intelligent Business Campus*
IIP –	*Intelligent Industrial Park*
IoE -	*Internet of Everything*
IoT -	*Internet of Things*
IP –	*Intellectual Property*
IPTV –	*Internet Protocol Television*
IREC –	*Intelligent Retail, Entertainment, & Convention center Complex*
IRR –	*Internal Rate of Return*
IT –	*Information Technology*
ITT –	*International Telephone & Telegraph*
JCL –	*Job Control Language*
MBA –	*Master's in Business Administration*
MPH -	*Miles Per Hour*
MRI –	*Magnetic Resonance Imaging*
MTBF –	*Mean Time Between Failures*
MTTF –	*Mean Time To Failures*
MTTR –	*Mean Time To Repair*
NBA –	*National Basketball Association*
NFC –	*Near Field Communications Chip*
NFL –	*National Football League*

NLR – *National Lambda Rail (high-speed fiber network)*

NPV – *Net Present Value*

OC - *Ongoing Costs (external)*

OIC - *Ongoing Internal Costs*

PBX – *Private Branch Exchange*

PC – *Personal Computer*

PSTN - *Public Switched Telephone Network*

R&D – *Research & Development*

REF – *Regional Economic Framework*

RFI – *Radio Frequency Interference*

RFI – *Request For Information*

RFID – *Radio Frequency Identification Chip*

RFP – *Request For Proposal*

RFQ – *Request For Quote*

RITE - *Real Estate, Infrastructure, Technology & Economic Development*

SLA – *Service Level Agreement*

SONET – *Synchronous Optical Network*

STS – *Shared Tenant Services*

Three Rs – *Rote, Repetition, & Routine*

TARGET -   *Technology And Revolutionary Gadgets Eventually Timeout (Map of Technology)*

TCI –   *Total Continuous Improvement*

TIF –   *Tax Increment Financing*

TQM -   *Total Quality Management*

TTC –   *True Total Costs*

VDM –   *Vortex of Declining Mediocrity*

VIP -   *Very Important Person*

Index

ABOUT THE AUTHOR

JAMES CARLINI

A long-time management consultant with a distinctive career spanning from Bell Telephone Laboratories, Motorola, Illinois Bell, and Arthur Young to his own consulting firm, he has worked on mission critical networks for the Bell System, 911 Centers, stock exchanges, banks, and other corporate networks. He has also been used as an expert witness in civil and federal courts, on cases ranging from network failures to wrongful deaths not to mention a U.S. Navy court-martial.

As a former award-winning faculty member for two decades at Northwestern University, a former elected official and state liquor commissioner, he draws from his experiences in many diverse roles to provide broad, practical insights into this convergence of these large areas. He is also the current Chairman of the Rolls-Royce Owners Club, Lake Michigan Chapter and member of the Bentley Drivers Club.

CPSIA information can be obtained at www.ICGtesting.com
Printed in the USA
LVOW01s1903181014

409222LV00001B/1/P